FREGE AND GÖDEL

FREGE AND GÖDEL

Two Fundamental Texts in Mathematical Logic

Edited by Jean van Heijenoort

toExcel
New York San Jose Lincoln Shanghai

Frege and Gödel

Two Fundamental Texts in Mathematical Logic

All Rights Reserved. Copyright © 1967, 1970, 1999
by the President and Fellows of Harvard College

No part of this book may be reproduced or transmitted in
any form or by any means, graphic, electronic, or mechanical,
including photocopying, recording, taping, or by any
information storage or retrieval system, without the
permission in writing from the publisher.

This edition republished by arrangement with toExcel,
an imprint of iUniverse.com, Inc.

For information address:
iUniverse.com, Inc.
620 North 48th Street
Suite 201
Lincoln, NE 68504-3467
www.iUniverse.com

ISBN: 1-58348-596-1

Printed in the United States of America

Preface

The present volume offers to the reader, in English translations, Frege's *Begriffsschrift* and Gödel's incompleteness paper. These two works make the respective years of their publication, 1879 and 1931, great years for logic. Frege's booklet brought to the world the theory of quantification and thus opened up a new epoch in the history of logic. Gödel's paper revealed intrinsic limitations of formal systems; by its results as well as its methods, by its direct impact and its indirect influence, the paper deeply marked the development of logic and foundations of mathematics. The reader thus has in his hands the two most important works in logic in modern times.

The contents of the present volume were originally published in *From Frege to Gödel: A source book in mathematical logic, 1879–1931*, edited by myself and published by Harvard University Press in 1967. In the present edition several misprints and minor errors have been corrected. The introductory notes that I wrote for the two papers are reproduced here. In these notes the reader will find, in particular, the credits for the translations.

JEAN VAN HEIJENOORT

Cambridge, Massachusetts
20 November 1969

Contents

Frege (*1879*). *Begriffsschrift*, a formula language, modeled upon that of arithmetic, for pure thought — 1

Gödel (*1930b*, *1931*, and *1931a*). Some metamathematical results on completeness and consistency, On formally undecidable propositions of *Principia mathematica* and related systems I, *and* On completeness and consistency — 83

References — 111

Index — 117

FREGE AND GÖDEL

Begriffsschrift, *a formula language, modeled upon that of arithmetic, for pure thought*

GOTTLOB FREGE

(*1879*)

This is the first work that Frege wrote in the field of logic, and, although a mere booklet of eighty-eight pages, it is perhaps the most important single work ever written in logic. Its fundamental contributions, among lesser points, are the truth-functional propositional calculus, the analysis of the proposition into function and argument(s) instead of subject and predicate, the theory of quantification, a system of logic in which derivations are carried out exclusively according to the form of the expressions, and a logical definition of the notion of mathematical sequence. Any single one of these achievements would suffice to secure the book a permanent place in the logician's library.

Frege was a mathematician by training;[a] the point of departure of his investigations in logic was a mathematical question, and mathematics left its mark upon his logical accomplishments. In studying the concept of number, Frege was confronted with difficulties when he attempted to give a logical analysis of the notion of sequence. The imprecision and ambiguity of ordinary language led him to look for a more appropriate tool; he devised a new mode of expression, a language that deals with the "conceptual content" and that he came to call "Begriffsschrift".[b] This ideography is a "formula language", that is, a *lingua characterica*, a language written with special symbols, "for pure thought", that is, free from rhetorical embellishments, "modeled upon that of arithmetic", that is, constructed from specific symbols that are manipulated according to definite rules. The last phrase does not mean that logic mimics arithmetic, and the analogies, uncovered by Boole and others, between logic and arithmetic are useless for Frege, precisely because he wants to employ logic in

---

[a] See his *Inaugural-Dissertation* (*1873*) and his thesis for *venia docendi* (*1874*).

[b] In the translation below this term is rendered by "ideography", a word used by Jourdain in a paper (*1912*) read and annotated by Frege; that Frege acquiesced in its use was the reason why ultimately it was adopted here. Another acceptable rendition is "concept writing", used by Austin (*Frege 1950*, p. 92e).

Professor Günther Patzig was so kind as to report in a private communication that a student of his, Miss Carmen Diaz, found an occurrence of the word "Begriffsschrift" in Trendelenburg (*1867*, p. 4, line 1), a work that Frege quotes in his preface to *Begriffsschrift* (see below, p. 6). Frege used the word in other writings, and in particular in his major work (*1893*, *1903*), but subsequently he seems to have become dissatisfied with it. In an unpublished fragment dated 26 July 1919 he writes: "I do not start from concepts in order to build up thoughts or propositions out of them; rather, I obtain the components of a thought by decomposition [[Zerfällung]] of the thought. In this respect my Begriffsschrift differs from the similar creations of Leibniz and his successors—in spite of its name, which perhaps I did not choose very aptly".

order to provide a foundation for arithmetic. He carefully keeps the logical symbols distinct from the arithmetic ones. Schröder (*1880*) criticized him for doing just that and thus wrecking a tradition established in the previous thirty years. Frege (*1882*, pp. 1–2) answered that his purpose had been quite different from that of Boole: "My intention was not to represent an abstract logic in formulas, but to express a content through written signs in a more precise and clear way than it is possible to do through words. In fact, what I wanted to create was not a mere *calculus ratiocinator* but a *lingua characterica* in Leibniz's sense".

Mathematics led Frege to an innovation that was to have a profound influence upon modern logic. He observes that we would do violence to mathematical statements if we were to impose upon them the distinction between subject and predicate. After a short but pertinent critique of that distinction, he replaces it by another, borrowed from mathematics but adapted to the needs of logic, that of function and argument. Frege begins his analysis by considering an ordinary sentence and remarks that the expression remains meaningful when certain words are replaced by others. A word for which we can make such successive substitutions occupies an argument place, and the stable component of the sentence is the function. This, of course, is not a definition, because in his system Frege deals not with ordinary sentences but with formulas; it is merely an explanation, after which he introduces functional letters and gives instructions for handling them and their arguments. Nowhere in the present text does Frege state what a function is or speak of the value of a function. He simply says that a judgment is obtained when the argument places between the parentheses attached to a functional letter have been properly filled (and, should the case so require, quantifiers have been properly used).

It is only in his subsequent writings (*1891* and thereafter) that Frege will devote a great deal of attention to the nature of a function.

Frege's booklet presents the propositional calculus in a version that uses the conditional and negation as primitive connectives. Other connectives are examined for a moment, and their intertranslatability with the conditional and negation is shown. Mostly to preserve the simple formulation of the rule of detachment, Frege decides to use these last two. The notation that he introduces for the conditional has often been criticized, and it has not survived. It presents difficulties in printing and takes up a large amount of space. But, as Frege himself (*1896*, p. 364) says, "the comfort of the typesetter is certainly not the *summum bonum*", and the notation undoubtedly allows one to perceive the structure of a formula at a glance and to perform substitutions with ease. Frege's definition of the conditional is purely truth-functional, and it leads him to the rule of detachment, stated in § 6. He notes the discrepancy between this truth-functional definition and ordinary uses of the word "if". Frege dismisses modal considerations from his logic with the remark that they concern the grounds for accepting a judgment, not the content of the judgment itself. Frege's use of the words "affirmed" and "denied", with his listing of all possible cases in the assignment of these terms to propositions, in fact amounts to the use of the truth-table method. His axioms for the propositional calculus (they are not independent) are formulas (1), (2), (8), (28), (31), and (41). His rules of inference are the rule of detachment and an unstated rule of substitution. A number of theorems of the propositional calculus are proved, but no question of completeness, consistency, or independence is raised.

Quantification theory is introduced in § 11. Frege's instructions how to use italic and German letters contain, in effect, the rule of generalization and the rule that allows us to infer $A \supset (x)F(x)$ from $A \supset F(x)$ when x does not occur free in A. There are three new axioms: (58) for instantiation, (52) and (54) for identity. No rule of substitution is explicitly stated, and one has to examine Frege's practice in his derivations to see what he allows. The substitutions are indicated by tables on the left of the derivations. These substitutions are simultaneous substitutions. When a substitution is specified with the help of "Γ", which plays the role of what we would today call a syntactic variable, particular care should be exercised, and it proves convenient to perform the substitutions that do not involve "Γ" before that involving "Γ" is carried out. The point will become clear to the reader if he compares, for example, the derivation of (51) with that of (98). Frege's derivations are quite detailed and, even in the absence of an explicit rule of substitution, can be unambiguously reconstructed.

Frege allows a functional letter to occur in a quantifier (p. 24 below). This license is not a necessary feature of quantification theory, but Frege has to admit it in his system for the definitions and derivations of the third part of the book. The result is that the difference between function and argument is blurred. In fact, even before coming to quantification over functions, Frege states (p. 24 below) that we can consider $\Phi(A)$ to be a function of the argument Φ as well as of the argument A. (This is precisely the point that Russell will seize upon to make it bear the brunt of his paradox—see *Russell 1902*). It is true that Frege writes (p. 24 below) that, if a functional letter occurs in a quantifier, "this circumstance must be taken into account". But the phrase remains vague. The most generous interpretation would be that, in the scope of the quantifier in which it occurs, a functional letter has to be treated as such, that is, must be provided with a pair of parentheses and one or more arguments. Frege, however, does not say as much, and in the derivation of formula (77) he substitutes $\mathfrak{F}$ for $\mathfrak{a}$ in $f(\mathfrak{a})$, at least as an intermediate step. If we also observe that in the derivation of formula (91) he substitutes $\mathfrak{F}$ for f, we see that he is on the brink of a paradox. He will fall into the abyss when (*1891*) he introduces the course-of-values of a function as something "complete in itself", which "may be taken as an **argument**". **For the rest of the story see** *van Heijenoort 1967*, pp. 124–128.

This flaw in Frege's system should not make us lose sight of the greatness of his achievement. The analysis of the proposition into function and argument, rather than subject and predicate, and quantification theory, which became possible only after such an analysis, are the very foundations of modern logic. The problems connected with quantification over functions could be approached only after a quantification theory had already been established. When the slowness and the wavering that marked the development of the propositional calculus are remembered, one cannot but marvel at seeing quantification theory suddenly coming full-grown into the world. Many years later (*1894*, p. 21) Peano still finds quantification theory "abstruse" and prefers to deal with it by means of just a few examples. Frege can proudly answer (*1896*, p. 376) that in 1879 he had already given all the laws of quantification theory; "these laws are few in **number, and I do not know why they should be said to be abstruse**".

In distinguishing his work from that of his predecessors and contemporaries, Frege repeatedly opposes a *lingua characterica* to a *calculus ratiocinator*. He uses these terms, suggested by Leibniz, to bring out an important feature of his system, in fact, one of the greatest achievements of his *Begriffsschrift*. In the pre-Fregean calculus of propositions and classes, logic, translated into formulas, is studied by means of arguments resting upon an intuitive logic. What Frege does is to construct logic as a language that need not be supplemented by any intuitive reasoning. Thus he is very careful to describe his system in purely *formal* terms (he even speaks of letters— Latin, German, and so on—rather than of variables, because of the imprecision of the latter term). He is fully aware that any system requires rules that cannot be expressed in the system; but these rules are void of any intuitive logic; they are "rules for the use of our signs" (p. 28 below): the rule of detachment, the rules for dealing with quantifiers. This is one of the great lessons of Frege's book. It was a new one in 1879, and it did not at once pervade the world of logic.

The third part of the book introduces a theory of mathematical sequences. Frege is moving toward his goal, the logical reconstruction of arithmetic. He defines the relation that Whitehead and Russell (*1910*, part II, sec. E) came to call the ancestral relation and that later (*1940*) Quine called the ancestral. The proper ancestral appears in § 26 and the ancestral proper in § 29. Subsequently Frege will use the notion for the justification of mathematical induction (*1884*, p. 93). Dedekind (*van Heijenoort 1967*, p. 101, and *1893*, XVII) recognized that the ancestral agrees in essence with his own notion of chain, which was publicly introduced nine years after Frege's notion.

At times *Begriffsschrift* begs for a clarification of linguistic usage, for a distinction between expressions and what these expressions refer to. In his subsequent writings Frege will devote a great deal of attention to this problem. On one point, however, the book touches upon them, and not too happily. In § 8 identity of content is introduced as a relation between names, not their contents. "⊢—$A = B$" means that the signs "A" and "B" have the same conceptual content and, according to Frege, is a statement about signs. There are strong arguments against such a conception, and Frege will soon recognize them. This will lead him to split the notion of conceptual content into sense ("Sinn") and reference ("Bedeutung") (*1892a*, but see also *1891*, p. 14; these two papers can be viewed as long emendations to *Begriffsschrift*).

In 1910 Jourdain sent to Frege the manuscript of a long paper that he had written on the history of logic and that contained a summary of *Begriffsschrift*. Frege answered with comments on a number of points, and Jourdain incorporated Frege's remarks in footnotes to his paper (*1912*). Some of these footnotes are reproduced below, at their appropriate places, with slight revisions in Jourdain's translation of Frege's comments (moreover, the German text used here is Frege's copy, and there are indications that the text that he sent to Jourdain and the copy that he preserved are not identical).

A few words should be said about Frege's use of the term "Verneinung". In a first use, "Verneinung" is opposed to "Bejahung", "verneinen" to "bejahen", and what these words express is, in fact, the ascription of truth values to contents of judgments; they are translated, respectively, by "denial" and "affirmation", "to deny" and "to affirm". The second use of "Verneinung" is for the connective, and when so used it is translated by "negation".

A number of misprints in the original were discovered during the translation. Most of them are included in the errata list that the reader will find in the reprint of Frege's booklet (*1964*, pp. 122–123).[o] Those that are not in that list are the following:

(1) On page XV, lines 6u, 5u, and 3u of the German text, "A" and "B" (which are alpha and beta) are not of the same font as "Φ" and "Ψ", while they should be;

---

[o] On misprints in *Frege 1964* see *Angelelli and Bynum 1966*.

(2) On page 29 of the German text, in § 15, the letters to the left of the long vertical line under (1) should be "a" and "b", not "a" and "b";

(3) The misprint indicated in footnote 18, p. 57 below;

(4) The misprint indicated in footnote 21, p. 65 below.

Moreover, Misprint 3 in the reprint's list does not occur in the German text used for the present translation; apparently, it is not a misprint at all but is simply due to the poor printing of some copies. The reprint also introduces misprints of its own: on page 1, line 4u, we find "——" where there should be "|——"; on page 62, near the top of the page, "$\frac{\gamma}{\beta}$" should be "$\frac{\gamma}{\beta}$."; on page 65 there should be a vertical negation stroke attached to the stroke preceding the first occurrence of "$h(y)$"; on page 39 an unreadable broken "c" has been left uncorrected.

The translation is by Stefan Bauer-Mengelberg, and it is published here by arrangement with Georg Olms Verlagsbuchhandlung.

PREFACE

In apprehending a scientific truth we pass, as a rule, through various degrees of certitude. Perhaps first conjectured on the basis of an insufficient number of particular cases, a general proposition comes to be more and more securely established by being connected with other truths through chains of inferences, whether consequences are derived from it that are confirmed in some other way or whether, conversely, it is seen to be a consequence of propositions already established. Hence we can inquire, on the one hand, how we have gradually arrived at a given proposition and, on the other, how we can finally provide it with the most secure foundation. The first question may have to be answered differently for different persons; the second is more definite, and the answer to it is connected with the inner nature of the proposition considered. The most reliable way of carrying out a proof, obviously, is to follow pure logic, a way that, disregarding the particular characteristics of objects, depends solely on those laws upon which all knowledge rests. Accordingly, we divide all truths that require justification into two kinds, those for which the proof can be carried out purely by means of logic and those for which it must be supported by facts of experience. But that a proposition is of the first kind is surely compatible with the fact that it could nevertheless not have come to consciousness in a human mind without any activity of the senses.[1] Hence it is not the psychological genesis but the best method of proof that is at the basis of the classification. Now, when I came to consider the question to which of these two kinds the judgments of arithmetic belong, I first had to ascertain how far one could proceed in arithmetic by means of inferences alone, with the sole support of those laws of thought that transcend all particulars. My initial step was to attempt to reduce the concept of ordering in a sequence to that of *logical* consequence, so as to proceed from there to the concept of number. To prevent anything intuitive [[Anschauliches]] from penetrating here unnoticed, I had to bend every effort to keep the chain of inferences free of gaps. In attempting to comply with this requirement in the strictest possible way I found the inadequacy of language to be an

---

[1] Since without sensory experience no mental development is possible in the beings known to us, that holds of all judgments.

obstacle; no matter how unwieldy the expressions I was ready to accept, I was less and less able, as the relations became more and more complex, to attain the precision that my purpose required. This deficiency led me to the idea of the present ideography. Its first purpose, therefore, is to provide us with the most reliable test of the validity of a chain of inferences and to point out every presupposition that tries to sneak in unnoticed, so that its origin can be investigated. That is why I decided to forgo expressing anything that is without significance for the *inferential sequence*. In § 3 I called what alone mattered to me the *conceptual content* [[*begrifflichen Inhalt*]]. Hence this definition must always be kept in mind if one wishes to gain a proper understanding of what my formula language is. That, too, is what led me to the name "Begriffsschrift". Since I confined myself for the time being to expressing relations that are independent of the particular characteristics of objects, I was also able to use the expression "formula language for pure thought". That it is modeled upon the formula language of arithmetic, as I indicated in the title, has to do with fundamental ideas rather than with details of execution. Any effort to create an artificial similarity by regarding a concept as the sum of its marks [[*Merkmale*]] was entirely alien to my thought. The most immediate point of contact between my formula language and that of arithmetic is the way in which letters are employed.

I believe that I can best make the relation of my ideography to ordinary language [[*Sprache des Lebens*]] clear if I compare it to that which the microscope has to the eye. Because of the range of its possible uses and the versatility with which it can adapt to the most diverse circumstances, the eye is far superior to the microscope. Considered as an optical instrument, to be sure, it exhibits many imperfections, which ordinarily remain unnoticed only on account of its intimate connection with our mental life. But, as soon as scientific goals demand great sharpness of resolution, the eye proves to be insufficient. The microscope, on the other hand, is perfectly suited to precisely such goals, but that is just why it is useless for all others.

This ideography, likewise, is a device invented for certain scientific purposes, and one must not condemn it because it is not suited to others. If it answers to these purposes in some degree, one should not mind the fact that there are no new truths in my work. I would console myself on this point with the realization that a development of method, too, furthers science. Bacon, after all, thought it better to invent a means by which everything could easily be discovered than to discover particular truths, and all great steps of scientific progress in recent times have had their origin in an improvement of method.

Leibniz, too, recognized—and perhaps overrated—the advantages of an adequate system of notation. His idea of a universal characteristic, of a *calculus philosophicus* or *ratiocinator*,[2] was so gigantic that the attempt to realize it could not go beyond the bare preliminaries. The enthusiasm that seized its originator when he contemplated the immense increase in the intellectual power of mankind that a system of notation directly appropriate to objects themselves would bring about led him to underestimate the difficulties that stand in the way of such an enterprise. But, even if this worthy goal cannot be reached in one leap, we need not despair of a slow, step-by-step approximation. When a problem appears to be unsolvable in its full generality, one should

---

[2] On that point see *Trendelenburg 1867* [[pp. 1–47, *Ueber Leibnizens Entwurf einer allgemeinen Charakteristik*]].

temporarily restrict it; perhaps it can then be conquered by a gradual advance. It is possible to view the signs of arithmetic, geometry, and chemistry as realizations, for specific fields, of Leibniz's idea. The ideography proposed here adds a new one to these fields, indeed the central one, which borders on all the others. If we take our departure from there, we can with the greatest expectation of success proceed to fill the gaps in the existing formula languages, connect their hitherto separated fields into a single domain, and extend this domain to include fields that up to now have lacked such a language.[3]

I am confident that my ideography can be successfully used wherever special value must be placed on the validity of proofs, as for example when the foundations of the differential and integral calculus are established.

It seems to me to be easier still to extend the domain of this formula language to include geometry. We would only have to add a few signs for the intuitive relations that occur there. In this way we would obtain a kind of *analysis situs*.

The transition to the pure theory of motion and then to mechanics and physics could follow at this point. The latter two fields, in which besides rational necessity 〚Denknothwendigkeit〛 empirical necessity 〚Naturnothwendigkeit〛 asserts itself, are the first for which we can predict a further development of the notation as knowledge progresses. That is no reason, however, for waiting until such progress appears to have become impossible.

If it is one of the tasks of philosophy to break the domination of the word over the human spirit by laying bare the misconceptions that through the use of language often almost unavoidably arise concerning the relations between concepts and by freeing thought from that with which only the means of expression of ordinary language, constituted as they are, saddle it, then my ideography, further developed for these purposes, can become a useful tool for the philosopher. To be sure, it too will fail to reproduce ideas in a pure form, and this is probably inevitable when ideas are represented by concrete means; but, on the one hand, we can restrict the discrepancies to those that are unavoidable and harmless, and, on the other, the fact that they are of a completely different kind from those peculiar to ordinary language already affords protection against the specific influence that a particular means of expression might exercise.

The mere invention of this ideography has, it seems to me, advanced logic. I hope that logicians, if they do not allow themselves to be frightened off by an initial impression of strangeness, will not withhold their assent from the innovations that, by a necessity inherent in the subject matter itself, I was driven to make. These deviations from what is traditional find their justification in the fact that logic has hitherto always followed ordinary language and grammar too closely. In particular, I believe that the replacement of the concepts *subject* and *predicate* by *argument* and *function*, respectively, will stand the test of time. It is easy to see how regarding a content as a function of an argument leads to the formation of concepts. Furthermore, the demonstration of the connection between the meanings of the words *if, and, not, or, there is, some, all,* and so forth, deserves attention.

Only the following point still requires special mention. The restriction, in § 6, to a

---

[3] 〚On that point see *Frege 1879a*.〛

single mode of inference is justified by the fact that, when the *foundations* for such an ideography are laid, the primitive components must be taken as simple as possible, if perspicuity and order are to be created. This does not preclude the possibility that *later* certain transitions from several judgments to a new one, transitions that this one mode of inference would not allow us to carry out except mediately, will be abbreviated into immediate ones. In fact this would be advisable in case of eventual application. In this way, then, further modes of inference would be created.

I noticed afterward that formulas (31) and (41) can be combined into a single one,

$$\vdash (\mathrm{\pi} a \equiv a),$$

which makes some further simplifications possible.

As I remarked at the beginning, arithmetic was the point of departure for the train of thought that led me to my ideography. And that is why I intend to apply it first of all to that science, attempting to provide a more detailed analysis of the concepts of arithmetic and a deeper foundation for its theorems. For the present I have reported in the third chapter some of the developments in this direction. To proceed farther along the path indicated, to elucidate the concepts of number, magnitude, and so forth—all this will be the object of further investigations, which I shall publish immediately after this booklet.

Jena, 18 December 1878.

CONTENTS

I. DEFINITION OF THE SYMBOLS

§ 1. Letters and other signs 10

Judgment

§ 2. Possibility that a content become a judgment. Content stroke, judgment stroke 11
§ 3. Subject and predicate. Conceptual content 12
§ 4. Universal, particular; negative; categoric, hypothetic, disjunctive; apodictic, assertory, problematic judgments 13

Conditionality

§ 5. If. Condition stroke 13
§ 6. Inference. The Aristotelian modes of inference 15

Negation

§ 7. Negation stroke. Or, either—or, and, but, and not, neither—nor 17

Identity of content

§ 8. Need for a sign for identity of content, introduction of such a sign....... 20

Functions

§ 9. Definition of the words "function" and "argument". Functions of several arguments. Argument places. Subject, object...................... 21

§ 10. Use of letters as function signs. "A has the property Φ." "B has the relation Ψ to A." "B is a result of an application of the procedure Ψ to the object A." The function sign as argument....................... 23

Generality

§ 11. German letters. The concavity in the content stroke. Replaceability of German letters. Their scope. Latin letters.......................... 24

§ 12. There are some objects that do not ——. There is no ——. There are some ——. Every. All. Causal connections. None. Some do not. Some. It is possible that ——. Square of logical opposition.................... 27

II. REPRESENTATION AND DERIVATION OF SOME JUDGMENTS OF PURE THOUGHT

§ 13. Usefulness of the deductive mode of presentation.................... 28
§ 14. The first two fundamental laws of conditionality.................... 29
§ 15. Some of their consequences....................................... 31
§ 16. The third fundamental law of conditionality, consequences........... 36
§ 17. The first fundamental law of negation, consequences................. 44
§ 18. The second fundamental law of negation, consequences............... 45
§ 19. The third fundamental law of negation, consequences................ 47
§ 20. The first fundamental law of identity of content, consequence........ 50
§ 21. The second fundamental law of identity of content, consequences...... 50
§ 22. The fundamental law of generality, consequences.................... 51

III. SOME TOPICS FROM A GENERAL THEORY OF SEQUENCES

§ 23. Introductory remarks... 55
§ 24. Heredity. Doubling of the judgment stroke. Lower-case Greek letters.... 55
§ 25. Consequences.. 57
§ 26. Succession in a sequence... 59
§ 27. Consequences.. 60
§ 28. Further consequences... 65
§ 29. "z belongs to the f-sequence beginning with x." Definition and consequences.. 69
§ 30. Further consequences... 71
§ 31. Single-valuedness of a procedure. Definition and consequences.......... 74

I. DEFINITION OF THE SYMBOLS

§ 1. The signs customarily employed in the general theory of magnitudes are of two kinds. The first consists of letters, of which each represents either a number left indeterminate or a function left indeterminate. This indeterminacy makes it possible to use letters to express the universal validity of propositions, as in

$$(a + b)c = ac + bc.$$

The other kind consists of signs such as $+$, $-$, $\sqrt{}$, 0, 1, and 2, of which each has its particular meaning.[4]

I adopt this basic idea of distinguishing two kinds of signs, which unfortunately is not strictly observed in the theory of magnitudes,[5] *in order to apply it in the more*

---

[4] ⟦Footnote by Jourdain (*1912*, p. 238):

Russell (*1908*) has expressed it: "A variable is a symbol which is to have one of a certain set of values, without its being decided which one. It does not have first one value of a set and then another; it has at all times *some* value of the set, where, so long as we do not replace the variable by a constant, the 'some' remains unspecified."

On the word "variable" Frege has supplied the note: "Would it not be well to omit this expression entirely, since it is hardly possible to define it properly? Russell's definition immediately raises the question what it means to say that 'a symbol has a value'. Is the relation of a sign to its significatum meant by this? In that case, however, we must insist that the sign be univocal, and the meaning (value) that the sign is to have must be determinate; then the variable would be a sign. But for him who does not subscribe to a formal theory a variable will not be a sign, any more than a number is. If, now, you write 'A variable is represented by a symbol that is to represent one of a certain set of values', the last defect is thereby removed; but what is the case then? The symbol represents, first, the variable and, second, a value taken from a certain supply without its being determined which. Accordingly, it seems better to leave the word 'symbol' out of the definition. The question as to what a variable is has to be answered independently of the question as to which symbol is to represent the variable. So we come to the definition: 'A variable is one of a certain set of values, without its being decided which one'. But the last addition does not yield any closer determination, and to belong to a certain set of values means, properly, to fall under a certain concept; for, after all, we can determine this set only by giving the properties that an object must have in order to belong to the set; that is, the set of values will be the extension of a concept. But, now, we can for every object specify a set of values to which it belongs, so that even the requirement that something is to be a value taken from a certain set does not determine anything. It is probably best to hold to the convention that Latin letters serve to confer generality of content on a theorem. And it is best not to use the expression 'variable' at all, since ultimately we cannot say either of a sign, or of what it expresses or denotes, that it is variable or that it is a variable, at least not in a sense that can be used in mathematics or logic. On the other hand, perhaps someone may insist that in '$(2 + x)(3 + x)$' the letter 'x' does not serve to confer generality of content on a proposition. But in the context of a proof such a formula will always occur as a part of a proposition, whether this proposition consists partly of words or exclusively of mathematical signs, and in such a context x will always serve to confer generality of content on a proposition. Now, it seems to me unfortunate to restrict to a particular set the values that are admissible for this letter. For we can always add the condition that a belong to this set, and then drop that condition. If an object Δ does not belong to the set, the condition is simply not satisfied and, if we replace 'a' by 'Δ' in the entire proposition, we obtain a true proposition. I would not say of a letter that it has a signification, a sense, a meaning, if it serves to confer generality of content on a proposition. We can replace the letter by the proper name 'Δ' of an object Δ; but this Δ cannot anyhow be regarded as the *meaning* of the letter; for it is not more closely allied with the letter than is any other object. Also, generality cannot be regarded as the meaning of the Latin letter; for it cannot be regarded as something independent, something that would be added to a content already complete in other respects. I would not, then, say 'terms whose meaning is indeterminate' or 'signs have variable meanings'. In this case signs have no denotations at all." [Frege, 1910.]⟧

[5] Consider 1, log, sin, lim.

comprehensive domain of pure thought in general. I therefore divide all signs that I use into *those by which we may understand different objects* and *those that have a completely determinate meaning.* The former are *letters* and they will serve chiefly to express *generality.* But, no matter how indeterminate the meaning of a letter, we must insist that throughout a given context the letter *retain* the meaning once given to it.

Judgment

§ 2. A judgment will always be expressed by means of the sign

$$\vdash ,$$

which stands to the left of the sign, or the combination of signs, indicating the content of the judgment. If we *omit* the small vertical stroke at the left end of the horizontal one, the judgment will be transformed into a *mere combination of ideas* [*Vorstellungsverbindung*],[6] of which the writer does not state whether he acknowledges it to be true or not. For example, let

$$\vdash\!\!\!-\!\!\!-A$$

stand for [[bedeute]] the judgment "Opposite magnetic poles attract each other";[7] then

$$-\!\!\!-\!\!\!-A$$

will not express [[ausdrücken]] this judgment;[8] it is to produce in the reader merely the idea of the mutual attraction of opposite magnetic poles, say in order to derive consequences from it and to test by means of these whether the thought is correct. When the vertical stroke is omitted, we express ourselves *paraphrastically*, using the words "the circumstance that" or "the proposition that".[9]

Not every content becomes a judgment when $\vdash\!\!\!-\!\!\!-$ is written before its sign; for

---

[6] [[Footnote by Jourdain (*1912*, p. 242):

"For this word I now simply say 'Gedanke'. The word 'Vorstellungsinhalt' is used now in a psychological, now in a logical sense. Since this creates obscurities, I think it best not to use this word at all in logic. We must be able to express a thought without affirming that it is true. If we want to characterize a thought as false, we must first express it without affirming it, then negate it, and affirm as true the thought thus obtained. We cannot correctly express a hypothetical connection between thoughts at all if we cannot express thoughts without affirming them, for in the hypothetical connection neither the thought appearing as antecedent nor that appearing as consequent is affirmed." [Frege, 1910.]]]

[7] I use Greek letters as abbreviations, and to each of these letters the reader should attach an appropriate meaning when I do not expressly give them a definition. [[The "*A*" that Frege is now using is a capital alpha.]]

[8] [[Jourdain had originally translated "bedeuten" by "signify", and Frege wrote (see *Jourdain 1912*, p. 242):

"Here we must notice the words 'signify' and 'express'. The former seems to correspond to 'bezeichnen' or 'bedeuten', the latter to 'ausdrucken'. According to the way of speaking I adopted I say 'A proposition expresses a thought and signifies its truth value'. Of a judgment we cannot properly say either that it signifies or that it is expressed. We do, to be sure, have a thought in the judgment, and that can be expressed; but we have more, namely, the recognition of the truth of this thought."]]

[9] [[Footnote by Jourdain (*1912*, p. 243):

"Instead of 'circumstance' and 'proposition' I would simply say 'thought'. Instead of 'beurtheilbarer Inhalt' we can also say 'Gedanke'." [Frege, 1910.]]]

example, the idea "house" does not. We therefore distinguish contents that *can become a judgment* from those that *cannot*.[10]

The horizontal stroke that is part of the sign ⊢—— *combines the signs that follow it into a totality, and the affirmation expressed by the vertical stroke at the left end of the horizontal one refers to this totality.* Let us call the horizontal stroke the *content stroke* and the vertical stroke the *judgment stroke*. The content stroke will in general serve to relate any sign to the totality of the signs that follow the stroke. *Whatever follows the content stroke must have a content that can become a judgment.*

§ 3. A distinction between *subject* and *predicate* does *not occur* in my way of representing a judgment. In order to justify this I remark that the contents of two judgments may differ in two ways: either the consequences derivable from the first, when it is combined with certain other judgments, always follow also from the second, when it is combined with these same judgments, [[and conversely,]] or this is not the case. The two propositions "The Greeks defeated the Persians at Plataea" and "The Persians were defeated by the Greeks at Plataea" differ in the first way. Even if one can detect a slight difference in meaning, the agreement outweighs it. Now I call that part of the content that is the *same* in both the *conceptual content*. Since *it alone* is of significance for our ideography, we need not introduce any distinction between propositions having the same conceptual content. If one says of the subject that it "is the concept with which the judgment is concerned", this is equally true of the object. We can therefore only say that the subject "is the concept with which the judgment is chiefly concerned". In ordinary language, the place of the subject in the sequence of words has the significance of a *distinguished* place, where we put that to which we wish especially to direct the attention of the listener (see also § 9). This may, for example, have the purpose of pointing out a certain relation of the given judgment to others and thereby making it easier for the listener to grasp the entire context. Now, all those peculiarities of ordinary language that result only from the interaction of speaker and listener—as when, for example, the speaker takes the expectations of the listener into account and seeks to put them on the right track even before the complete sentence is enunciated—have nothing that answers to them in my formula language, since in a judgment I consider only that which influences its *possible consequences*. Everything necessary for a correct inference is expressed in full, but what is not necessary is generally not indicated; *nothing is left to guesswork.* In this I faithfully follow the example of the formula language of mathematics, a language to which one would do violence if he were to distinguish between subject and predicate in it. We can imagine a language in which the proposition "Archimedes perished at the capture of Syracuse" would be expressed thus: "The violent death of Archimedes at the capture of Syracuse is a fact". To be sure, one can distinguish between subject and predicate here, too, if one wishes to do so, but the subject contains the whole content, and the predicate serves only to turn the content into a judgment. *Such a*

---

[10] On the other hand, the circumstance that there are houses, or that there is a house (see § 12 [[footnote 15]]), is a content that can become a judgment. But the idea "house" is only a part of it. In the proposition "The house of Priam was made of wood" we could not put "circumstance that there is a house" in place of "house". For a different kind of example of a content that cannot become a judgment see the passage following formula (81).

[[In German Frege's distinction is between "beurtheilbare" and "unbeurtheilbare" contents. Jourdain uses the words "judicable" and "nonjudicable".]]

language would have only a single predicate for all judgments, namely, "*is a fact*". We see that there cannot be any question here of subject and predicate in the ordinary sense. *Our ideography is a language of this sort, and in it the sign* ⊢—— *is the common predicate for all judgments.*

In the first draft of my formula language I allowed myself to be misled by the example of ordinary language into constructing judgments out of subject and predicate. But I soon became convinced that this was an obstacle to my specific goal and led only to useless prolixity.

§ 4. The remarks that follow are intended to explain the significance for our purposes of the distinctions that we introduce among judgments.

We distinguish between *universal* and *particular* judgments; this is really not a distinction between judgments but between contents. *We ought to say "a judgment with a universal content", "a judgment with a particular content"*. For these properties hold of the content even when it is *not* advanced as a judgment but as a ⟦mere⟧ proposition (see § 2).

The same holds of negation. In an indirect proof we say, for example, "Suppose that the line segments AB and CD are not equal". Here the content, that the line segments AB and CD are not equal, contains a negation; but this content, though it can become a judgment, is nevertheless not advanced as a judgment. Hence the negation attaches to the content, whether this content becomes a judgment or not. I therefore regard it as more appropriate to consider negation as an adjunct of a *content that can become a judgment*.

The distinction between categoric, hypothetic, and disjunctive judgments seems to me to have only grammatical significance.[11]

The apodictic judgment differs from the assertory in that it suggests the existence of universal judgments from which the proposition can be inferred, while in the case of the assertory one such a suggestion is lacking. By saying that a proposition is necessary I give a hint about the grounds for my judgment. *But, since this does not affect the conceptual content of the judgment, the form of the apodictic judgment has no significance for us.*

If a proposition is advanced as possible, either the speaker is suspending judgment by suggesting that he knows no laws from which the negation of the proposition would follow or he says that the generalization of this negation is false. In the latter case we have what is usually called a *particular affirmative judgment* (see § 12). "It is possible that the earth will at some time collide with another heavenly body" is an instance of the first kind, and "A cold can result in death" of the second.

Conditionality

§ 5. If A and B stand for contents that can become judgments (§ 2), there are the following four possibilities:

(1) A is affirmed and B is affirmed;
(2) A is affirmed and B is denied;
(3) A is denied and B is affirmed;
(4) A is denied and B is denied.

[11] The reason for this will be apparent from the entire book.

Now

$$\vdash\!\!\!\begin{array}{c}\rule{1em}{0.4pt}A\\\rule{1em}{0.4pt}B\end{array}$$

stands for the judgment that *the third of these possibilities does not take place, but one of the three others does.* Accordingly, if

$$\begin{array}{c}\rule{1em}{0.4pt}A\\\rule{1em}{0.4pt}B\end{array}$$

is denied, this means that the third possibility takes place, hence that A is denied and B affirmed.

Of the cases in which

$$\begin{array}{c}\rule{1em}{0.4pt}A\\\rule{1em}{0.4pt}B\end{array}$$

is affirmed we single out for comment the following three:

(1) A must be affirmed. Then the content of B is completely immaterial. For example, let $\vdash\!\!\!\rule{1em}{0.4pt}A$ stand for $3 \times 7 = 21$ and B for the circumstance that the sun is shining. Then only the first two of the four cases mentioned are possible. There need not exist a causal connection between the two contents.

(2) B has to be denied. Then the content of A is immaterial. For example, let B stand for the circumstance that perpetual motion is possible and A for the circumstance that the world is infinite. Then only the second and fourth of the four cases are possible. There need not exist a causal connection between A and B.

(3) We can make the judgment

$$\vdash\!\!\!\begin{array}{c}\rule{1em}{0.4pt}A\\\rule{1em}{0.4pt}B\end{array}$$

without knowing whether A and B are to be affirmed or denied. For example, let B stand for the circumstance that the moon is in quadrature with the sun and A for the circumstance that the moon appears as a semicircle. In that case we can translate

$$\vdash\!\!\!\begin{array}{c}\rule{1em}{0.4pt}A\\\rule{1em}{0.4pt}B\end{array}$$

by means of the conjunction "if": "If the moon is in quadrature with the sun, the moon appears as a semicircle". The causal connection inherent in the word "if", however, is not expressed by our signs, even though only such a connection can provide the ground for a judgment of the kind under consideration. For causal connection is something general, and we have not yet come to express generality (see § 12).

Let us call the vertical stroke connecting the two horizontal ones the *condition stroke*. The part of the upper horizontal stroke to the left of the condition stroke is the content stroke for the meaning, just explained, of the combination of signs

$$\begin{array}{c}\rule{1em}{0.4pt}A\\\rule{1em}{0.4pt}B\end{array};$$

to it is affixed any sign that is intended to relate to the total content of the expression. The part of the horizontal stroke between A and the condition stroke is the content stroke of A. The horizontal stroke to the left of B is the content stroke of B. Accordingly, it is easy to see that

$$\vdash \begin{array}{c} A \\ B \\ \Gamma \end{array}$$

denies the case in which A is denied and B and Γ are affirmed. We must think of this as having been constructed from

$$\begin{array}{c} A \\ B \end{array}$$

and Γ in the same way as

$$\begin{array}{c} A \\ B \end{array}$$

was constructed from A and B. We therefore first have the denial of the case in which

$$\begin{array}{c} A \\ B \end{array}$$

is denied and Γ is affirmed. But the denial of

$$\begin{array}{c} A \\ B \end{array}$$

means that A is denied and B is affirmed. From this we obtain what was given above. If a causal connection is present, we can also say "A is the necessary consequence of B and Γ", or "If the circumstances B and Γ occur, then A also occurs".

It is no less easy to see that

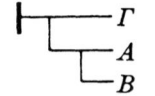

$$\vdash \begin{array}{c} \Gamma \\ A \\ B \end{array}$$

denies the case in which B is affirmed but A and Γ are denied.[12] If we assume that there exists a causal connection between A and B, we can translate the formula as "If A is a necessary consequence of B, one can infer that Γ takes place".

§ 6. The definition given in § 5 makes it apparent that from the two judgments

$$\vdash \begin{array}{c} A \\ B \end{array} \quad \text{and} \quad \vdash B$$

the new judgment

$$\vdash A$$

---

[12] ⟦There is an oversight here, already pointed out by Schröder (*1880*, p. 88).⟧

follows. Of the four cases enumerated above, the third is excluded by

$$\vdash\!\!\begin{array}{c}A\\B\end{array}$$

and the second and fourth by

$$\vdash\!\!-B,$$

so that only the first remains.

We could write this inference perhaps as follows:

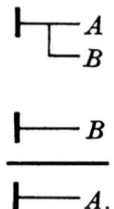

$$\vdash\!\!\begin{array}{c}A\\B\end{array}$$

$$\vdash\!\!-B$$
$$\overline{\vdash\!\!-A.}$$

This would become awkward if long expressions were to take the places of A and B, since each of them would have to be written twice. That is why I use the following abbreviation. To every judgment occurring in the context of a proof I assign a number, which I write to the right of the judgment at its first occurrence. Now assume, for example, that the judgment

$$\vdash\!\!\begin{array}{c}A\\B,\end{array}$$

or one containing it as a special case, has been assigned the number X. Then I write the inference as follows:

$$(\text{X}): \quad \dfrac{\vdash\!\!-B}{\vdash\!\!-A.}$$

Here it is left to the reader to put the judgment

$$\vdash\!\!\begin{array}{c}A\\B\end{array}$$

together for himself from $\vdash\!\!-B$ and $\vdash\!\!-A$ and to see whether he obtains the judgment X that has been invoked or a special case thereof.

If, for example, the judgment $\vdash\!\!-B$ has been assigned the number XX, I also write the same inference as follows:

$$(\text{XX})::\quad \dfrac{\vdash\!\!\begin{array}{c}A\\B\end{array}}{\vdash\!\!-A.}$$

Here the double colon indicates that $\vdash\!\!-B$, which was only referred to by XX, would have to be formed, from the two judgments written down, in a way different from that above.

Furthermore if, say, the judgment ⊢——Γ had been assigned the number XXX, I would abbreviate the two judgments

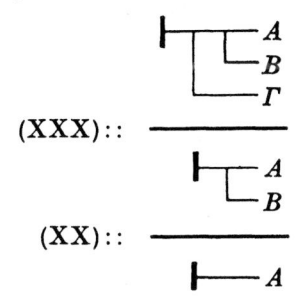

$$(XXX) :: \quad \dfrac{\vdash\!\!\begin{array}{l} -A \\ -B \\ -\Gamma \end{array}}{\vdash\!\!\begin{array}{l} -A \\ -B \end{array}}$$

$$(XX) :: \quad \dfrac{\vdash\!\!\begin{array}{l} -A \\ -B \end{array}}{\vdash\!\!-A}$$

still more thus:

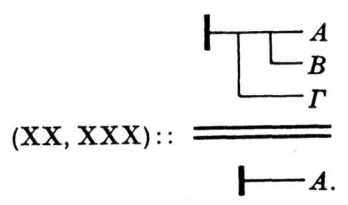

$$(XX, XXX) :: \quad \dfrac{\vdash\!\!\begin{array}{l} -A \\ -B \\ -\Gamma \end{array}}{\vdash\!\!-A.}$$

Following Aristotle, we can enumerate quite a few modes of inference in logic; I employ only this one, at least in all cases in which a new judgment is derived from more than a single one. For, the truth contained in some other kind of inference can be stated in one judgment, of the form: if M holds and if N holds, then Λ holds also, or, in signs,

$$\vdash\!\!\begin{array}{l} -\Lambda \\ -M \\ -N. \end{array}$$

From this judgment, together with ⊢——N and ⊢——M, there follows, as above, ⊢——Λ. In this way an inference in accordance with any mode of inference can be reduced to our case. Since it is therefore possible to manage with a single mode of inference, it is a commandment of perspicuity to do so. Otherwise there would be no reason to stop at the Aristotelian modes of inference; instead, one could continue to add new ones indefinitely: from each of the judgments expressed in a formula in §§ 13–22 we could make a particular mode of inference. *With this restriction to a single mode of inference, however, we do not intend in any way to state a psychological proposition; we wish only to decide a question of form in the most expedient way.* Some of the judgments that take the place of Aristotelian kinds of inference will be listed in § 22 (formulas (59), (62), and (65)).

Negation

§ 7. If a short vertical stroke is attached below the content stroke, this will express the circumstance that *the content does not take place*. So, for example,

$$\vdash_{\top}\!\!-A$$

means "A does not take place". I call this short vertical stroke the *negation stroke*.

The part of the horizontal stroke to the right of the negation stroke is the content stroke of A; the part to the left of the negation stroke is the content stroke of the negation of A. If there is no judgment stroke, then here—as in any other place where the ideography is used—no judgment is made.

$$\mathbf{-}\!\!\top\!\!-A$$

merely calls upon us to form the idea that A does not take place, without expressing whether this idea is true.

We now consider some cases in which the signs of conditionality and negation are combined.

$$\vdash\!\!\top\!\!\!\begin{array}{l}\top\!\!-A\\ \mathrel{\rule{0pt}{1ex}\smash{\rule[0.5ex]{0pt}{0pt}}}\!-B\end{array}$$

means "The case in which B is to be affirmed and the negation of A to be denied does not take place"; in other words, "The possibility of affirming both A and B does not exist", or "A and B exclude each other". Thus only the following three cases remain:

A is affirmed and B is denied;
A is denied and B is affirmed;
A is denied and B is denied.

In view of the preceding it is easy to state what the significance of each of the three parts of the horizontal stroke to the left of A is.

$$\vdash\!\!\begin{array}{l}-A\\ \top\!\!-B\end{array}$$

means "The case in which A is denied and the negation of B is affirmed does not obtain", or "A and B cannot both be denied". Only the following possibilities remain:

A is affirmed and B is affirmed;
A is affirmed and B is denied;
A is denied and B is affirmed;

A and B together exhaust all possibilities. Now the words "or" and "either—or" are used in two ways: "A or B" means, in the first place, just the same as

$$\begin{array}{l}-A\\ \top\!\!-B,\end{array}$$

hence it means that no possibility other than A and B is thinkable. For example, if a mass of gas is heated, its volume or its pressure increases. In the second place, the expression "A or B" combines the meanings of both

$$\begin{array}{l}\top\!\!-A\\ -B\end{array}\quad\text{and}\quad\begin{array}{l}-A\\ \top\!\!-B,\end{array}$$

so that no third is possible besides A and B, and, moreover, that A and B exclude each other. Of the four possibilities, then, only the following two remain:

A is affirmed and B is denied;
A is denied and B is affirmed.

Of the two ways in which the expression "A or B" is used, the first, which does not exclude the coexistence of A and B, is the more important, and *we shall use the word "or" in this sense.* Perhaps it is appropriate to distinguish between "or" and "either—or" by stipulating that only the latter shall have the secondary meaning of mutual exclusion. We can then translate

$$\begin{array}{l} \top A \\ \bot B \end{array}$$

by "A or B". Similarly,

$$\begin{array}{l} \top A \\ \top B \\ \top \varGamma \end{array}$$

has the meaning of "A or B or $\varGamma$".

$$\vdash \begin{array}{l} \top A \\ \bot B \end{array}$$

means

"$\begin{array}{l} \top A \\ \bot B \end{array}$ is denied",

or "The case in which both A and B are affirmed occurs". The three possibilities that remained open for

$$\begin{array}{l} \top A \\ \bot B \end{array}$$

are, however, excluded. Accordingly, we can translate

$$\vdash \begin{array}{l} \top A \\ \bot B \end{array}$$

by "Both A and B are facts". It is also easy to see that

$$\begin{array}{l} \top A \\ \bot B \\ \top \varGamma \end{array}$$

can be rendered by "A and B and $\varGamma$". If we want to represent in signs "Either A or B" with the secondary meaning of mutual exclusion, we must express

"$\begin{array}{l}\top A \\ \bot B\end{array}$ and $\begin{array}{l}- A \\ \top B\end{array}$."

This yields

$$\begin{array}{l}\top A \\ \bot B \\ \top A \\ \top B\end{array} \text{ or also } \begin{array}{l}\top A \\ \top B \\ \top A \\ \bot B\end{array}.$$

Instead of expressing the "and", as we did here, by means of the signs of conditionality and negation, we could on the other hand also represent conditionality by means of a sign for "and" and the sign of negation. We could introduce, say,

$$\left\{\begin{array}{c}\Gamma\\\Delta\end{array}\right.$$

as a sign for the total content of Γ and Δ, and then render

$$\vdash\!\!\begin{array}{c}A\\B\end{array}$$

by

$$\left\{\begin{array}{c}\top A\\ \\B.\end{array}\right.$$

I chose the other way because I felt that it enables us to express inferences more simply. The distinction between "and" and "but" is of the kind that is not expressed in the present ideography. The speaker uses "but" when he wants to hint that what follows is different from what one might at first expect.

$$\vdash\!\!\begin{array}{c}A\\B\end{array}$$

means "Of the four possibilities the third, namely, that A is denied and B is affirmed, occurs". We can therefore translate it as "B takes place and (but) A does not".

We can translate the combination of signs

$$\vdash\!\!\begin{array}{c}B\\A\end{array}$$

by the same words.

$$\vdash\!\!\begin{array}{c}B\\A\end{array}$$

means "The case in which both A and B are denied occurs". Hence we can translate it as "Neither A nor B is a fact". What has been said here about the words "or", "and", and "neither —nor" applies, of course, only when they connect contents that *can become judgments*.

Identity of content

§ 8. Identity of content differs from conditionality and negation in that it applies to names and not to contents. Whereas in other contexts signs are merely representatives of their content, so that every combination into which they enter expresses only a relation between their respective contents, they suddenly display their own selves when they are combined by means of the sign for identity of content; for it expresses the circumstance that two names have the same content. Hence the introduction of a sign for identity of content necessarily produces a bifurcation in the meaning of all signs: they stand at times for their content, at times for themselves. At first we have the impression that what we are dealing with pertains merely to the *expression* and *not to the thought*, that we do not need different signs at all for the same content and hence no sign whatsoever for identity of content. To show that this is an empty illusion I take the following example from geometry. Assume that on the circumference of a circle there is a fixed point A about which a ray revolves. When this ray passes through the center of the circle, we call the other point at which it intersects the circle the point B associated with this position of the ray. The point of intersection, other than A, of the ray and the circumference will then be called the point B associated with the position of the ray at any time; this point is such that continuous variations in its position must always correspond to continuous variations in the position of the ray. Hence the name B denotes something indeterminate so long as the corresponding position of the ray has not been specified. We can now ask: what point is associated with the position of the ray when it is perpendicular to the diameter? The answer will be: the point A. In this case, therefore, the name B has the same content as has the name A; and yet we could not have used only one name from the beginning, since the justification for that is given only by the answer. One point is determined in two ways: (1) immediately through intuition and (2) as a point B associated with the ray perpendicular to the diameter.

To each of these ways of determining the point there corresponds a particular name. Hence the need for a sign for identity of content rests upon the following consideration: the same content can be completely determined in different ways; but that in a particular case *two ways of determining it* really yield the *same result* is the content of a *judgment*. Before this judgment can be made, two distinct names, corresponding to the two ways of determining the content, must be assigned to what these ways determine. The judgment, however, requires for its expression a sign for identity of content, a sign that connects these two names. From this it follows that the existence of different names for the same content is not always merely an irrelevant question of form; rather, that there are such names is the very heart of the matter if each is associated with a different way of determining the content. In that case the judgment that has the identity of content as its object is synthetic, in the Kantian sense. A more extrinsic reason for the introduction of a sign for identity of content is that it is at times expedient to introduce an abbreviation for a lengthy expression. Then we must express the identity of content that obtains between the abbreviation and the original form.

Now let

$$\vdash (A \equiv B)$$

mean that *the sign A and the sign B have the same conceptual content, so that we can everywhere put B for A and conversely.*

Functions

§ 9. Let us assume that the circumstance that hydrogen is lighter than carbon dioxide is expressed in our formula language; we can then replace the sign for hydrogen by the sign for oxygen or that for nitrogen. This changes the meaning in such a way that "oxygen" or "nitrogen" enters into the relations in which "hydrogen" stood before. If we imagine that an expression can thus be altered, it decomposes into a stable component, representing the totality of relations, and the sign, regarded as replaceable by others, that denotes the object standing in these relations. The former component I call a function, the latter its argument. The distinction has nothing to do with the conceptual content; it comes about only because we view the expression in a particular way. According to the conception sketched above, "hydrogen" is the argument and "being lighter than carbon dioxide" the function; but we can also conceive of the same conceptual content in such a way that "carbon dioxide" becomes the argument and "being heavier than hydrogen" the function. We then need only regard "carbon dioxide" as replaceable by other ideas, such as "hydrochloric acid" or "ammonia".

"The circumstance that carbon dioxide is heavier than hydrogen" and "The circumstance that carbon dioxide is heavier than oxygen" are the same function with different arguments if we regard "hydrogen" and "oxygen" as arguments; on the other hand, they are different functions of the same argument if we regard "carbon dioxide" as the argument.

To consider another example, take "The circumstance that the center of mass of the solar system has no acceleration if internal forces alone act on the solar system". Here "solar system" occurs in two places. Hence we can consider this as a function of the argument "solar system" in various ways, according as we think of "solar system" as replaceable by something else at its first occurrence, at its second, or at both (but then in both places by the same thing). These three functions are all different. The situation is the same for the proposition that Cato killed Cato. If we here think of "Cato" as replaceable at its first occurrence, "to kill Cato" is the function; if we think of "Cato" as replaceable at its second occurrence, "to be killed by Cato" is the function; if, finally, we think of "Cato" as replaceable at both occurrences, "to kill oneself" is the function.

We now express the matter generally.

If in an expression, whose content need not be capable of becoming a judgment, a simple or a compound sign has one or more occurrences and if we regard that sign as replaceable in all or some of these occurrences by something else (but everywhere by the same thing), then we call the part that remains invariant in the expression a function, and the replaceable part the argument of the function.

Since, accordingly, something can be an argument and also occur in the function at places where it is not considered replaceable, we distinguish in the function between the argument places and the others.

Let us warn here against a false impression that is very easily occasioned by linguistic usage. If we compare the two propositions "The number 20 can be represented as the sum of four squares" and "Every positive integer can be represented as the sum of four squares", it seems to be possible to regard "being representable as the sum of four squares" as a function that in one case has the argument "the number 20" and in the other "every positive integer". We see that this view is mistaken if we observe that "the number 20" and "every positive integer" are not concepts of the same rank [gleichen Ranges]. What is asserted of the number 20 cannot be asserted in the same sense of "every positive integer", though under certain circumstances it can be asserted of every positive integer. The expression "every positive integer" does not, as does "the number 20", by itself yield an independent idea but acquires a meaning only from the context of the sentence.

For us the fact that there are various ways in which the same conceptual content can be regarded as a function of this or that argument has no importance so long as function and argument are completely determinate. But, if the argument becomes *indeterminate*, as in the judgment "You can take as argument of 'being representable as the sum of four squares' an arbitrary positive integer, and the proposition will always be true", then the distinction between function and argument takes on a *substantive* [[*inhaltliche*]] significance. On the other hand, it may also be that the argument is determinate and the function indeterminate. In both cases, through the opposition between the *determinate* and the *indeterminate* or that between the *more* and the *less* determinate, the whole is decomposed into *function* and *argument* according to its content and not merely according to the point of view adopted.

If, given a function, we think of a sign[13] *that was hitherto regarded as not replaceable as being replaceable at some or all of its occurrences, then by adopting this conception we obtain a function that has a new argument in addition to those it had before.* This procedure yields *functions of two or more arguments*. So, for example, "The circumstance that hydrogen is lighter than carbon dioxide" can be regarded as function of the two arguments "hydrogen" and "carbon dioxide".

In the mind of the speaker the subject is ordinarily the main argument; the next in importance often appears as object. Through the choice between [[grammatical]] forms, such as active—passive, or between words, such as "heavier"—"lighter" and "give"—"receive", ordinary language is free to allow this or that component of the sentence to appear as main argument at will, a freedom that, however, is restricted by the scarcity of words.

§ 10. *In order to express an indeterminate function of the argument A, we write A, enclosed in parentheses, to the right of a letter*, for example

$$\Phi(A).$$

Likewise,

$$\Psi(A, B)$$

means a function of the two arguments A and B that is not determined any further. Here the occurrences of A and B in the parentheses represent the occurrences of A and B in the function, irrespective of whether these are single or multiple for A or for B. Hence in general

$$\Psi(A, B)$$

differs from

$$\Psi(B, A).$$

Indeterminate functions of more arguments are expressed in a corresponding way. We can read

$$\vdash\!\!\!-\!\!\!- \Phi(A)$$

---

[13] We can now regard a sign that previously was considered replaceable [[in some places]] as replaceable also in those places in which up to this point it was considered fixed.

as "A has the property Φ".

$$\vdash \Psi(A, B)$$

can be translated by "B stands in the relation Ψ to A" or "B is a result of an application of the procedure Ψ to the object A".

Since the sign Φ occurs in the expression $\Phi(A)$ and since we can imagine that it is replaced by other signs, Ψ or X, which would then express other functions of the argument A, *we can also regard $\Phi(A)$ as a function of the argument Φ*. This shows quite clearly that the concept of function in analysis, which in general I used as a guide, is far more restricted than the one developed here.

Generality

§ 11. In the expression of a judgment we can always regard the combination of signs to the right of $\vdash$ as a function of one of the signs occurring in it. *If we replace this argument by a German letter and if in the content stroke we introduce a concavity with this German letter in it, as in*

$$\vdash \overset{\mathfrak{a}}{\smile} \Phi(\mathfrak{a}),$$

this stands for the judgment that, whatever we may take for its argument, the function is a fact. Since a letter used as a sign for a function, such as Φ in $\Phi(A)$, can itself be regarded as the argument of a function, its place can be taken, in the manner just specified, by a German letter. The meaning of a German letter is subject only to the obvious restrictions that, if a combination of signs following a content stroke can become a judgment (§ 2), this possibility remain unaffected by such a replacement and that, if the German letter occurs as a function sign, this circumstance be taken into account. *All other conditions to be imposed on what may be put in place of a German letter are to be incorporated into the judgment.* From such a judgment, therefore, we can always derive an arbitrary number of *judgments of less general content* by substituting each time something else for the German letter and then removing the concavity in the content stroke. The horizontal stroke to the left of the concavity in

$$\vdash \overset{\mathfrak{a}}{\smile} \Phi(\mathfrak{a})$$

is the content stroke for the circumstance that, whatever we may put in place of $\mathfrak{a}$, $\Phi(\mathfrak{a})$ holds; the horizontal stroke to the right of the concavity is the content stroke of $\Phi(\mathfrak{a})$, and here we must imagine that something definite has been substituted for $\mathfrak{a}$.

According to what we said above about the significance of the judgment stroke, it is easy to see what an expression like

$$\overset{\mathfrak{a}}{\smile} X(\mathfrak{a})$$

means. It can occur as a part of a judgment, like

$$\vdash \top \overset{\mathfrak{a}}{\smile} X(\mathfrak{a}) \quad \text{or} \quad \vdash \begin{array}{c} \phantom{\overset{\mathfrak{a}}{\smile}} A \\ \overset{\mathfrak{a}}{\smile} X(\mathfrak{a}). \end{array}$$

It is clear that from these judgments we cannot derive less general judgments by substituting something definite for $\mathfrak{a}$, as we could from

$$\vdash\!\!\!-\!\!\stackrel{\mathfrak{a}}{\smile}\!\!-\Phi(\mathfrak{a}).$$

$\vdash\!\!\!-\!\!\stackrel{\mathfrak{a}}{\smile}\!\!-X(\mathfrak{a})$ denies that, whatever we may put in place of $\mathfrak{a}$, $X(\mathfrak{a})$ is always a fact. This does not by any means deny that we could specify some meaning Δ for $\mathfrak{a}$ such that $X(\Delta)$ would be a fact.

$$\vdash\!\!\!-\!\!\begin{array}{c}\rule{1em}{0.4pt}A\\ \rule{0.4pt}{1em}\!\!\stackrel{\mathfrak{a}}{\smile}\!\!-X(\mathfrak{a})\end{array}$$

means that the case in which $-\!\!\stackrel{\mathfrak{a}}{\smile}\!\!-X(\mathfrak{a})$ is affirmed and A is denied does not occur. But this does not by any means deny that the case in which $X(\Delta)$ is affirmed and A is denied does occur; for, as we just saw, $X(\Delta)$ can be affirmed and $-\!\!\stackrel{\mathfrak{a}}{\smile}\!\!-X(\mathfrak{a})$ can still be denied. Hence we cannot put something arbitrary in place of $\mathfrak{a}$ here either without endangering the truth of the judgment. This explains why the concavity with the German letter written into it is necessary: *it delimits the scope [[Gebiet]] that the generality indicated by the letter covers. The German letter retains a fixed meaning only within its own scope*; within one judgment the same German letter can occur in different scopes, without the meaning attributed to it in one scope extending to any other. The scope of a German letter can include that of another, as is shown by the example

$$\vdash\!\!\!-\!\!\stackrel{\mathfrak{a}}{}\!\!\begin{array}{c}\rule{1em}{0.4pt}A(\mathfrak{a})\\ \rule{0.4pt}{1em}\!\!\stackrel{\mathfrak{e}}{\smile}\!\!-B(\mathfrak{a},\mathfrak{e}).\end{array}$$

In that case they must be chosen *different*; we could not put $\mathfrak{a}$ for $\mathfrak{e}$. Replacing a German letter everywhere in its scope by some other one is, of course, permitted, so long as in places where different letters initially stood different ones also stand afterward. This has no effect on the content. *Other substitutions are permitted only if the concavity immediately follows the judgment stroke*, that is, if the content of the entire judgment constitutes the scope of the German letter. Since, accordingly, that case is a distinguished one, I shall introduce the following abbreviation for it. *An [[italic]] Latin letter always is to have as its scope the content of the entire judgment*, and this fact need not be indicated by a concavity in the content stroke. If a Latin letter occurs in an expression that is not preceded by a judgment stroke, the expression is meaningless. *A Latin letter may always be replaced by a German one that does not yet occur in the judgment*; then the concavity must be introduced immediately following the judgment stroke. For example, instead of

$$\vdash\!\!\!-\!\!-X(a)$$

we can write

$$\vdash\!\!\!-\!\!\stackrel{\mathfrak{a}}{\smile}\!\!-X(\mathfrak{a})$$

if a occurs only in the argument places of $X(a)$.

It is clear also that from

$$\vdash\!\!\begin{array}{l}\!\!-\Phi(a)\\\!\!-A\end{array}$$

we can derive

$$\vdash\!\!-\underset{a}{\smile}\!\!-\!\!\begin{array}{l}\Phi(\mathfrak{a})\\ A\end{array}$$

if A is an expression in which $\mathfrak{a}$ does not occur and if $\mathfrak{a}$ stands only in the argument places of $\Phi(\mathfrak{a})$.[14] If $-\underset{a}{\smile}-\Phi(\mathfrak{a})$ is denied, we must be able to specify a meaning for $\mathfrak{a}$ such that $\Phi(\mathfrak{a})$ will be denied. If, therefore, $-\underset{a}{\smile}-\Phi(\mathfrak{a})$ were to be denied and A to be affirmed, we would have to be able to specify a meaning for $\mathfrak{a}$ such that A would be affirmed and $\Phi(\mathfrak{a})$ would be denied. But on account of

$$\vdash\!\!\begin{array}{l}\!\!-\Phi(a)\\\!\!-A\end{array}$$

we cannot do that; for this means that, whatever $\mathfrak{a}$ may be, the case in which $\Phi(\mathfrak{a})$ is denied and A is affirmed is excluded. Therefore we cannot deny $-\underset{a}{\smile}-\Phi(\mathfrak{a})$ and affirm A; that is,

$$\vdash\!\!-\underset{a}{\smile}\!\!-\!\!\begin{array}{l}\Phi(\mathfrak{a})\\ A.\end{array}$$

Likewise, from

$$\vdash\!\!\begin{array}{l}\!\!-\Phi(a)\\\!\!-A\\\!\!-B\end{array}$$

we can deduce

$$\vdash\!\!-\underset{a}{\smile}\!\!-\!\!\begin{array}{l}\Phi(\mathfrak{a})\\ A\\ B\end{array}$$

if $\mathfrak{a}$ does not occur in A or B and $\Phi(a)$ contains a only in the argument places. This case can be reduced to the preceding one, since

$$\vdash\!\!\begin{array}{l}\!\!-\Phi(a)\\\!\!-A\\\!\!-B\end{array}$$

can be written

$$\vdash\!\!\begin{array}{l}\!\!-\Phi(a)\\\!\!-A\\\!\!-B\end{array}$$

---

[14] 〚Footnote by Jourdain (*1912*, p. 248):
Frege remarked [Frege, 1910] that "it is correct that one can give up the distinguishing use of Latin, German, and perhaps also of Greek letters, but at the cost of perspicuity of formulas".〛

and since we can transform

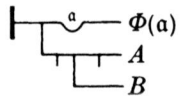

$$\vdash \smile^{\mathfrak{a}} \Phi(\mathfrak{a}),\ A,\ B$$

back into

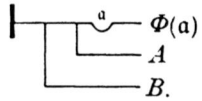

$$\vdash \smile^{\mathfrak{a}} \Phi(\mathfrak{a}),\ A,\ B.$$

Similar considerations apply when still more condition strokes are present.

§ 12. We now consider certain combinations of signs.

$$\vdash \smile^{\mathfrak{a}} X(\mathfrak{a})$$

means that we could find some object, say Δ, such that $X(\Delta)$ would be denied. We can therefore translate it as "There are some objects that do not have property X".

The meaning of

$$\vdash \smile^{\mathfrak{a}}\top X(\mathfrak{a})$$

differs from this. The formula means "Whatever $\mathfrak{a}$ may be, $X(\mathfrak{a})$ must always be denied", or "There does not exist anything having property X", or, if we call something that has property X an X, "There is no X".

$$\smile^{\mathfrak{a}}\top \Lambda(\mathfrak{a})$$

is denied by

$$\vdash \smile^{\mathfrak{a}}\top \Lambda(\mathfrak{a}).$$

We can therefore translate the last formula as "There are Λ".[15]

$$\vdash \smile^{\mathfrak{a}}\top P(\mathfrak{a}),\ X(\mathfrak{a})$$

means "Whatever we may put in place of $\mathfrak{a}$, the case in which $P(\mathfrak{a})$ would have to be denied and $X(\mathfrak{a})$ to be affirmed does not occur". Thus it is possible here that, for some meanings that can be given to $\mathfrak{a}$, $P(\mathfrak{a})$ would have to be affirmed and $X(\mathfrak{a})$ to be affirmed, for others $P(\mathfrak{a})$ would have to be affirmed and $X(\mathfrak{a})$ to be denied, and for others still $P(\mathfrak{a})$ would have to be denied and $X(\mathfrak{a})$ to be denied. We could therefore translate it as "If something has property X, it also has property P", "Every X is a P", or "All X are P".

This is the way in which causal connections are expressed.

$$\vdash \smile^{\mathfrak{a}}\top P(\mathfrak{a}),\ \Psi(\mathfrak{a})$$

---

[15] This must be understood in such a way as to include the case "There exists one Λ" as well. If, for example, $\Lambda(x)$ means the circumstance that x is a house, then

$$\vdash \smile^{\mathfrak{a}}\top \Lambda(\mathfrak{a})$$

reads "There are houses or there is at least one house". See footnote 10.

means "No meaning can be given to $\mathfrak{a}$ such that both $P(\mathfrak{a})$ and $\Psi(\mathfrak{a})$ could be affirmed". We can therefore translate it as "What has property Ψ does not have property P" or "No Ψ is a P".

$$\vdash \smile^{\mathfrak{a}} \biggl[\begin{array}{l} P(\mathfrak{a}) \\ \Lambda(\mathfrak{a}) \end{array}$$

denies

$$\smile^{\mathfrak{a}} \biggl[\begin{array}{l} P(\mathfrak{a}) \\ \Lambda(\mathfrak{a}) \end{array}$$

and can therefore be rendered by "Some Λ are not P".

$$\vdash \smile^{\mathfrak{a}} \top \biggl[\begin{array}{l} P(\mathfrak{a}) \\ M(\mathfrak{a}) \end{array}$$

denies that no M is a P and therefore means "Some[16] M are P", or "It is possible that a M be a P".

Thus we obtain the square of logical opposition:

$$\smile^{\mathfrak{a}} \biggl[\begin{array}{l} P(\mathfrak{a}) \\ X(\mathfrak{a}) \end{array} \quad \text{contrary} \quad \smile^{\mathfrak{a}} \top \biggl[\begin{array}{l} P(\mathfrak{a}) \\ X(\mathfrak{a}) \end{array}$$

subalternate *contradictory* *contradictory* subalternate

$$\top \smile^{\mathfrak{a}} \top \biggl[\begin{array}{l} P(\mathfrak{a}) \\ X(\mathfrak{a}) \end{array} \quad \text{[[sub]]contrary} \quad \top \smile^{\mathfrak{a}} \biggl[\begin{array}{l} P(\mathfrak{a}) \\ X(\mathfrak{a}) \end{array}$$

II. REPRESENTATION AND DERIVATION OF SOME JUDGMENTS OF PURE THOUGHT

§ 13. We have already introduced a number of fundamental principles of thought in the first chapter in order to transform them into rules for the use of our signs. These rules and the laws whose transforms they are cannot be expressed in the ideography because they form its basis. Now in the present chapter a number of judgments of pure thought for which this is possible will be represented in signs. It seems natural to derive the more complex of these judgments from simpler ones, not in order to make them more certain, which would be unnecessary in most cases, but in order to make manifest the relations of the judgments to one another. Merely to know the laws is obviously not the same as to know them together with the connections that

---

[16] The word "some" must always be understood here in such a way as to include the case "one" as well. More explicitly we would say "some or at least one".

some have to others. In this way we arrive at a small number of laws in which, if we add those contained in the rules, the content of all the laws is included, albeit in an undeveloped state. And that the deductive mode of presentation makes us acquainted with that core is another of its advantages. Since in view of the boundless multitude of laws that can be enunciated we cannot list them all, we cannot achieve completeness except by searching out those that, *by their power*, contain all of them. Now it must be admitted, certainly, that the way followed here is not the only one in which the reduction can be done. That is why not all relations between the laws of thought are elucidated by means of the present mode of presentation. There is perhaps another set of judgments from which, when those contained in the rules are added, all laws of thought could likewise be deduced. Still, with the method of reduction presented here such a multitude of relations is exhibited that any other derivation will be much facilitated thereby.

The propositions forming the core of the presentation below are nine in number. To express three of these, formulas (1), (2), and (8), we require besides letters only the sign of conditionality; formulas (28), (31), and (41) contain in addition the sign of negation; two, formulas (52) and (54), contain that of identity of content; and in one, formula (58), the concavity in the content stroke is used.

The derivations that follow would tire the reader if he were to retrace them in every detail; they serve merely to insure that the answer to any question concerning the derivation of a law is at hand.

§ 14. 
$$\begin{array}{c} a \\ b \\ a \end{array} \tag{1}$$

says "The case in which a is denied, b is affirmed, and a is affirmed is excluded". This is evident, since a cannot at the same time be denied and affirmed. We can also express the judgment in words thus, "If a proposition a holds, then it also holds in case an arbitrary proposition b holds". Let a, for example, stand for the proposition that the sum of the angles of the triangle ABC is two right angles, and b for the proposition that the angle ABC is a right angle. Then we obtain the judgment "If the sum of the angles of the triangle ABC is two right angles, this also holds in case the angle ABC is a right angle".

The (1) to the right of

$$\begin{array}{c} a \\ b \\ a \end{array}$$

is the number of this formula.

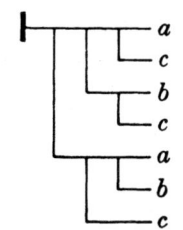

$$\begin{array}{c} a \\ c \\ b \\ c \\ a \\ b \\ c \end{array} \tag{2}$$

means "The case in which

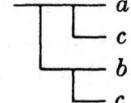

$$\begin{array}{l}\!\!\!\!-\!\!\!\begin{array}{l}-a\\ \llcorner c\end{array}\\ \llcorner\begin{array}{l}-b\\ \llcorner c\end{array}\end{array}$$

is denied and

$$\begin{array}{l}\!\!\!\!-\!\!\!\begin{array}{l}-a\\ \llcorner b\end{array}\\ \llcorner c\end{array}$$

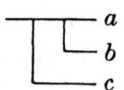

is affirmed does not take place".
   But

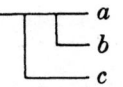

$$\begin{array}{l}\!\!\!\!-\!\!\!\begin{array}{l}-a\\ \llcorner b\end{array}\\ \llcorner c\end{array}$$

means the circumstance that the case in which a is denied, b is affirmed, and c is affirmed is excluded. The denial of

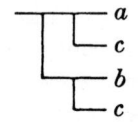

$$\begin{array}{l}\!\!\!\!-\!\!\!\begin{array}{l}-a\\ \llcorner c\end{array}\\ \llcorner\begin{array}{l}-b\\ \llcorner c\end{array}\end{array}$$

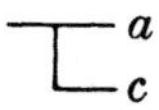

says that $\begin{array}{l}-a\\ \llcorner c\end{array}$ is denied and $\begin{array}{l}-b\\ \llcorner c\end{array}$ is affirmed. But the denial of $\begin{array}{l}-a\\ \llcorner c\end{array}$ means that a is denied and c is affirmed. Thus the denial of

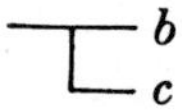

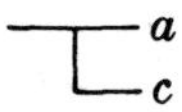

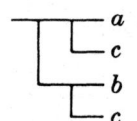

$$\begin{array}{l}\!\!\!\!-\!\!\!\begin{array}{l}-a\\ \llcorner c\end{array}\\ \llcorner\begin{array}{l}-b\\ \llcorner c\end{array}\end{array}$$

means that a is denied, c is affirmed, and $\begin{array}{l}-b\\ \llcorner c\end{array}$ is affirmed. But the affirmation of $\begin{array}{l}-b\\ \llcorner c\end{array}$ and that of c entails the affirmation of b. That is why the denial of

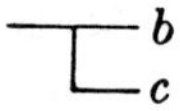

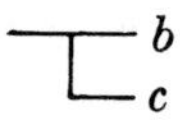

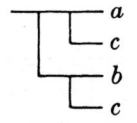

$$\begin{array}{l}\!\!\!\!-\!\!\!\begin{array}{l}-a\\ \llcorner c\end{array}\\ \llcorner\begin{array}{l}-b\\ \llcorner c\end{array}\end{array}$$

has as a consequence the denial of a and the affirmation of b and c. Precisely this case is excluded by the affirmation of

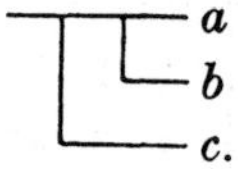

$$\begin{array}{l}\!\!\!\!-\!\!\!\begin{array}{l}-a\\ \llcorner b\end{array}\\ \llcorner c.\end{array}$$

Thus the case in which

$$\begin{array}{l}\!\!\!\!-\!\!\!\begin{array}{l}-a\\ \llcorner b\end{array}\\ \llcorner\begin{array}{l}-c\\ \llcorner b\end{array}\end{array}$$

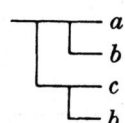

is denied and

$$\vdash\!\!\!\begin{array}{l}a\\b\\c\end{array}$$

is affirmed cannot take place, and that is what the judgment

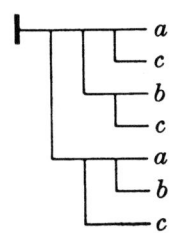

$$\vdash\!\!\!\begin{array}{l}a\\c\\b\\c\\a\\b\\c\end{array}$$

asserts. For the case in which causal connections are present, we can also express this as follows: "If a proposition a is a necessary consequence of two propositions b and c, that is, if

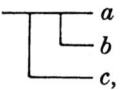

$$\begin{array}{l}a\\b\\c,\end{array}$$

and if one of these, b, is in turn a necessary consequence of the other, c, then the proposition a is a necessary consequence of this latter one, c, alone".

For example, let c mean that in a sequence Z of numbers every successor term is greater than its predecessor, let b mean that a term M is greater than L, and let a mean that the term N is greater than L. Then we obtain the following judgment: "If from the propositions that in the number sequence Z every successor term is greater than its predecessor and that the term M is greater than L it can be inferred that the term N is greater than L, and if from the proposition that in the number sequence Z every successor term is greater than its predecessor it follows that M is greater than L, then the proposition that N is greater than L can be inferred from the proposition that every successor term in the number sequence Z is greater than its predecessor".

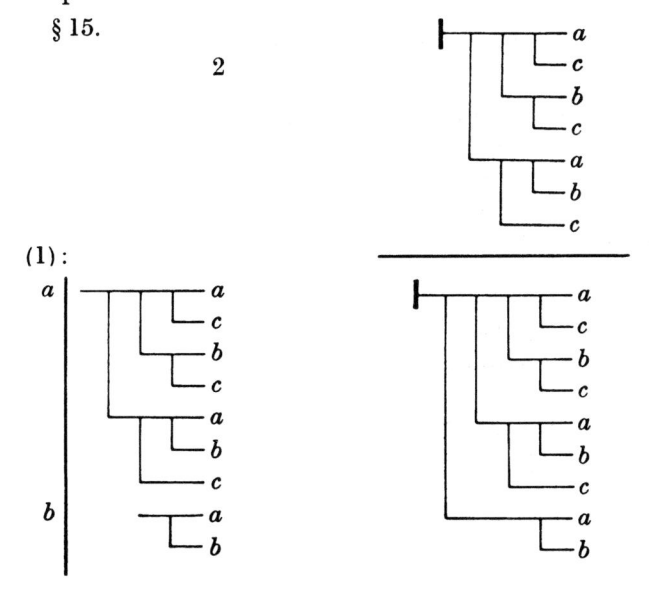

§ 15.

2

$$\vdash\!\!\!\begin{array}{l}a\\c\\b\\c\\a\\b\\c\end{array}$$

(1):

(3).

The 2 on the left indicates that formula (2) stands to its right. The inference that brings about the transition from (2) and (1) to (3) is expressed by an abbreviation in accordance with § 6. In full it would be written as follows:

1

2

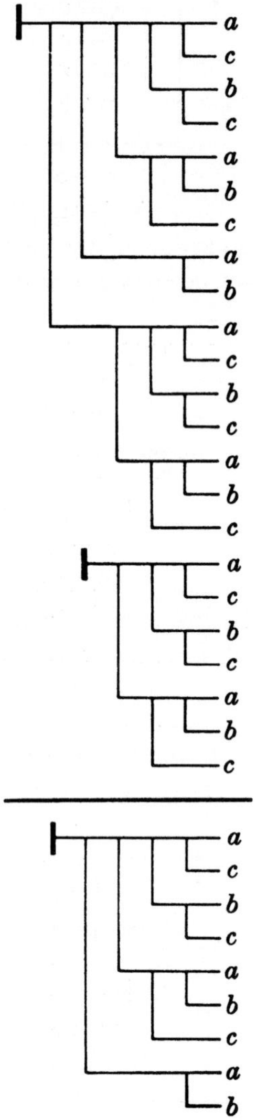

(3).

The small table under the (1) serves to make proposition (1) more easily recognizable in the more complicated form it takes here. It states that in

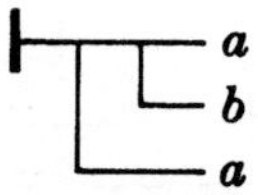

BEGRIFFSSCHRIFT

we are to put

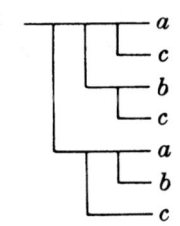

$$\vdash\!\!\begin{array}{l} a \\ c \\ b \\ c \\ a \\ b \\ c \end{array}$$

in place of a and

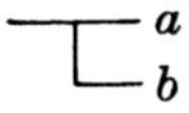

$$\vdash\!\!\begin{array}{l} a \\ b \end{array}$$

in place of b.

3

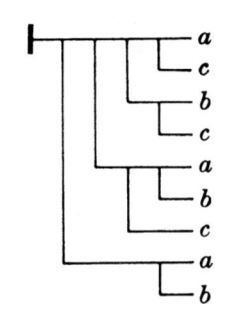

$$\vdash\!\!\begin{array}{l} a \\ c \\ b \\ c \\ a \\ b \\ c \\ a \\ b \end{array}$$

(2):

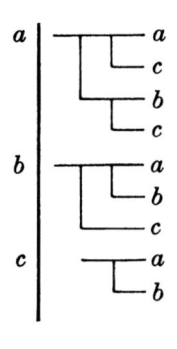

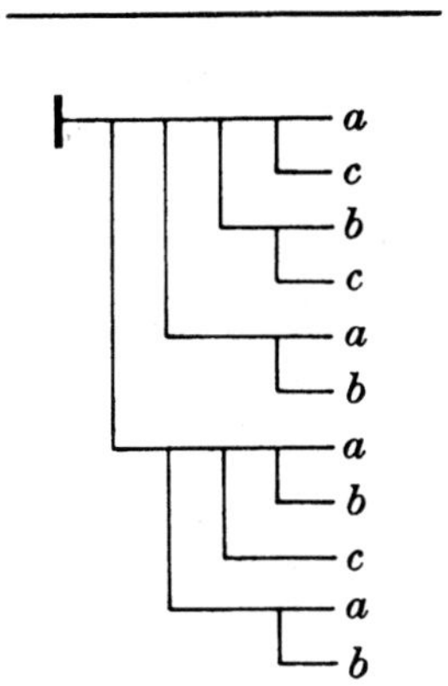

a	$\begin{array}{l} a \\ c \\ b \\ c \end{array}$
b	$\begin{array}{l} a \\ b \\ c \end{array}$
c	$\begin{array}{l} a \\ b \end{array}$

$$\vdash\!\!\begin{array}{l} a \\ c \\ b \\ c \\ a \\ b \\ a \\ b \\ c \\ a \\ b \end{array} \qquad (4).$$

The table under the (2) means that in

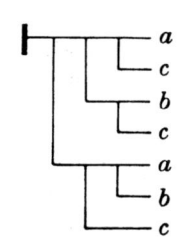

$$\vdash\!\!\begin{array}{l} a \\ c \\ b \\ c \\ a \\ b \\ c \end{array}$$

we are to put in place of a, b, and c, respectively, the expressions standing to the right of them; as a result we obtain

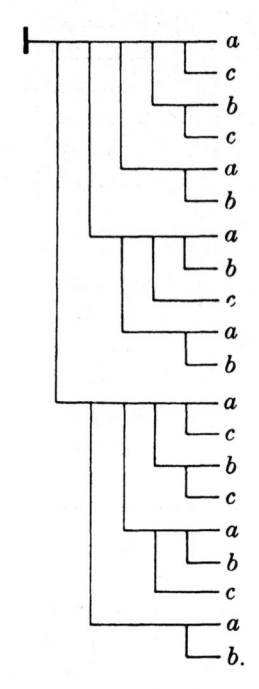

We readily see how (4) follows from this and (3).

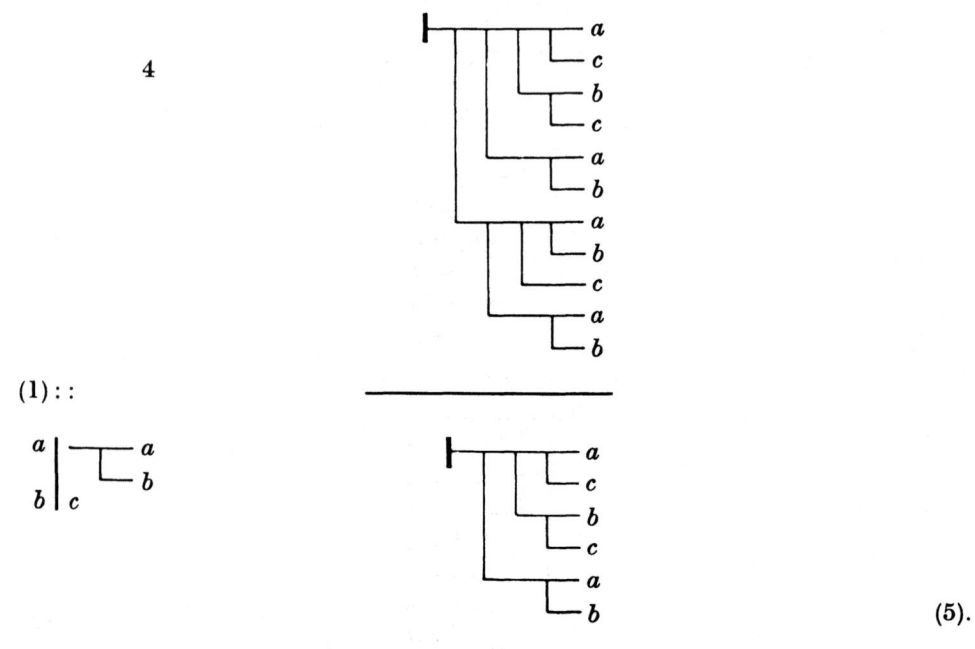

$$4$$

$$(1)::$$

(5).

The significance of the double colon is explained in § 6.

Example for (5). Let a be the circumstance that the piece of iron E becomes magnetized, b the circumstance that a galvanic current flows through the wire D, and c the circumstance that the key T is depressed. We then obtain the judgment: "If the proposition holds that E becomes magnetized as soon as a galvanic current flows through D and if the proposition holds that a galvanic current flows through D as soon as T is depressed, then E becomes magnetized if T is depressed".

If causal connections are assumed, (5) can be expressed thus: "If b is a sufficient condition for a and if c is a sufficient condition for b, then c is a sufficient condition for a".

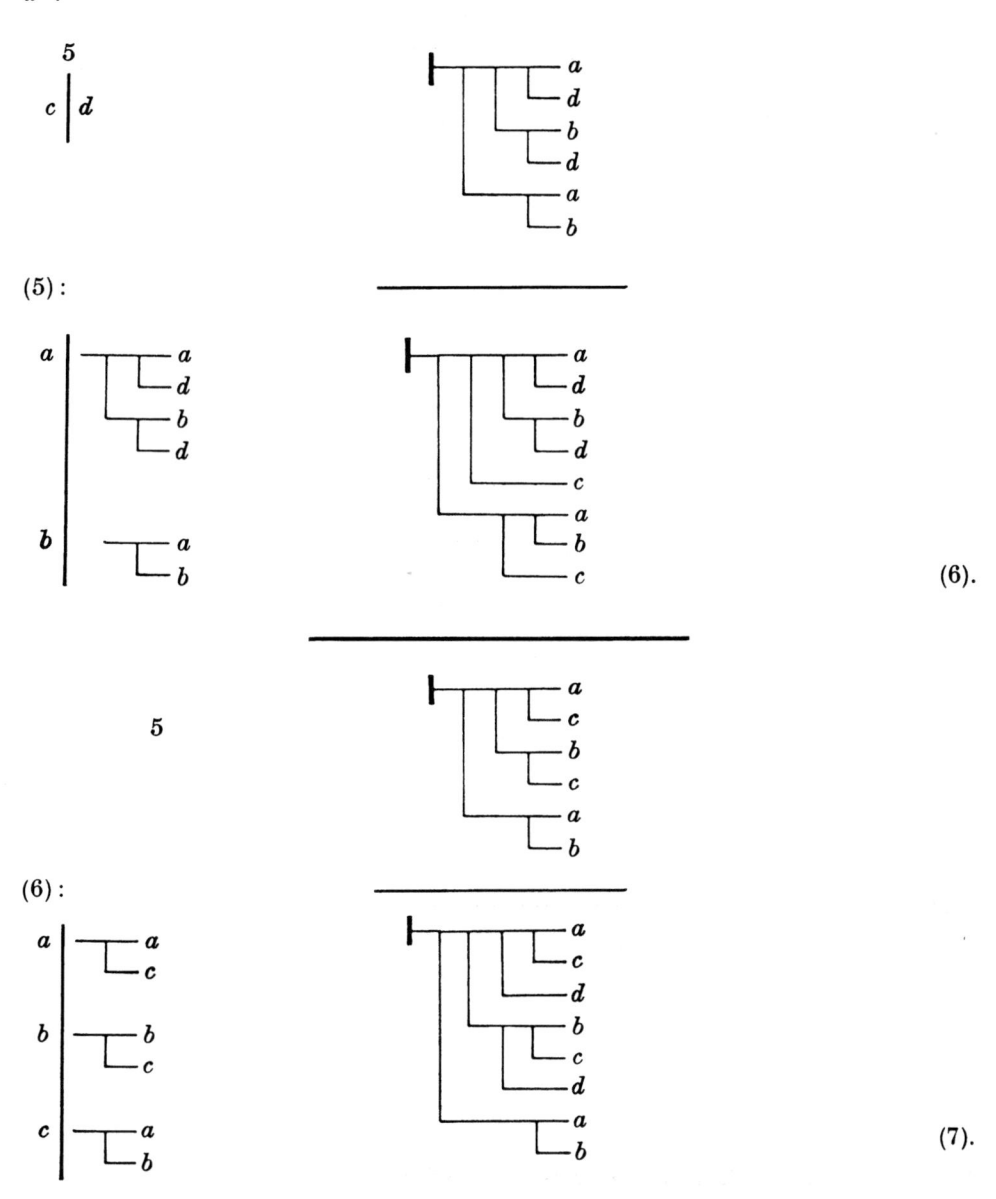

This proposition differs from (5) only in that instead of one condition, c, we now have two, c and d.

Example for (7). Let d mean the circumstance that the piston K of an air pump is moved from its leftmost position to its rightmost position, c the circumstance that the valve H is in position I, b the circumstance that the density D of the air in the cylinder of the air pump is reduced by half, and a the circumstance that the height H of a barometer connected to the inside of the cylinder decreases by half. Then we obtain the judgment: "If the proposition holds that the height H of the barometer decreases by half as soon as the density D of the air is reduced by half, and if the proposition holds that the density D of the air is reduced by half if the piston K is moved from the leftmost to the rightmost position and if the valve is in position I, then it follows that the height H of the barometer decreases by half if the piston K is moved from the leftmost to the rightmost position while the valve H is in position I".

§ 16.

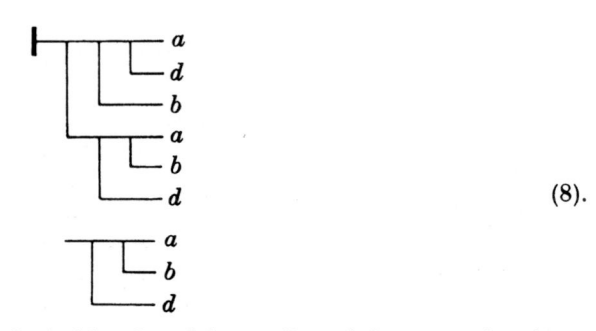

$$\vdash\!\!\begin{array}{c}\rule{1em}{0.4pt}\!\!\begin{array}{l}a\\d\end{array}\\b\\\rule{1em}{0.4pt}\!\!\begin{array}{l}a\\b\end{array}\\d\end{array} \qquad (8).$$

$$\begin{array}{c}\rule{1em}{0.4pt}\!\!\begin{array}{l}a\\b\end{array}\\d\end{array}$$

means that the case in which a is denied but b and d are affirmed does not take place;

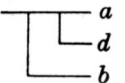

$$\begin{array}{c}\rule{1em}{0.4pt}\!\!\begin{array}{l}a\\d\end{array}\\b\end{array}$$

means the same, and (8) says that the case in which

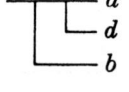

$$\begin{array}{c}\rule{1em}{0.4pt}\!\!\begin{array}{l}a\\d\end{array}\\b\end{array}$$

is denied and

$$\begin{array}{c}\rule{1em}{0.4pt}\!\!\begin{array}{l}a\\b\end{array}\\d\end{array}$$

is affirmed is excluded. This can also be expressed thus: "If two conditions have a proposition as a consequence, their order is immaterial".

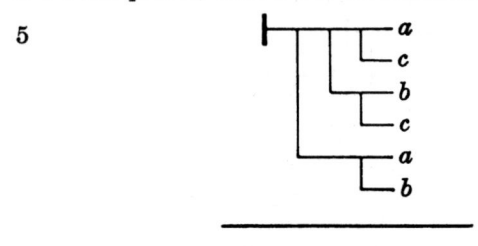

5

$$\vdash\!\!\begin{array}{c}\rule{1em}{0.4pt}\!\!\begin{array}{l}a\\c\end{array}\\b\\c\\\rule{1em}{0.4pt}\!\!\begin{array}{l}a\\b\end{array}\end{array}$$

(8): ──────────

$$\begin{array}{c|c} a & \vdash a \\ & \llcorner c \\ b & \vdash b \\ & \llcorner c \\ d & \vdash a \\ & \llcorner b \end{array} \qquad \vdash\!\!\!\begin{array}{l} a \\ \llcorner c \\ a \\ \llcorner b \\ b \\ \llcorner c \end{array} \qquad (9).$$

This proposition differs from (5) only in an unessential way.

$$\overset{8}{\begin{array}{c|c} a & b \\ b & e \end{array}} \qquad \vdash\!\!\!\begin{array}{l} b \\ \llcorner d \\ e \\ b \\ e \\ d \end{array}$$

$(9):$

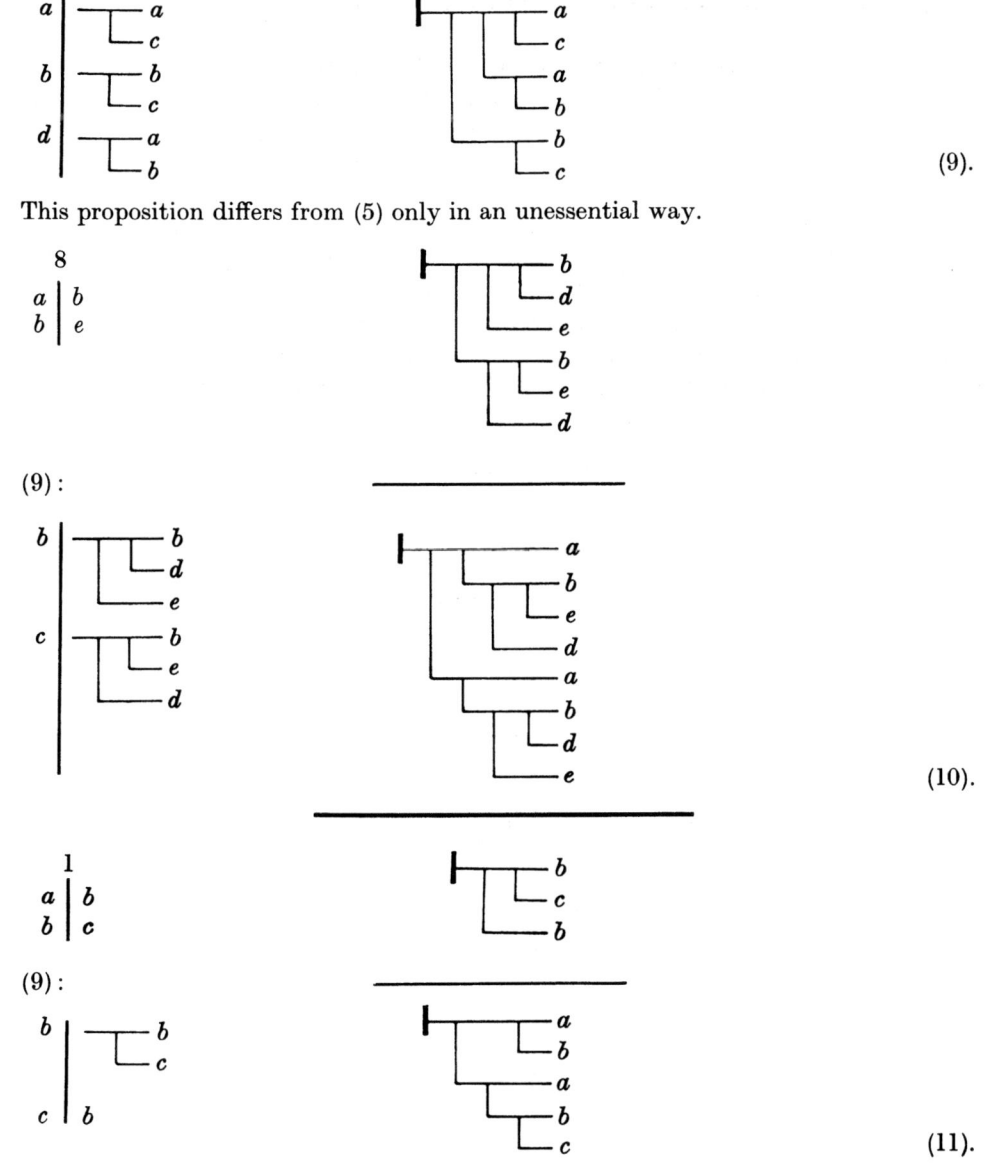

$$\begin{array}{c|c} b & \vdash b \\ & \llcorner d \\ & e \\ c & \vdash b \\ & \llcorner e \\ & d \end{array} \qquad \vdash\!\!\!\begin{array}{l} a \\ \llcorner b \\ e \\ d \\ a \\ b \\ d \\ e \end{array} \qquad (10).$$

$$\overset{1}{\begin{array}{c|c} a & b \\ b & c \end{array}} \qquad \vdash\!\!\!\begin{array}{l} b \\ \llcorner c \\ b \end{array}$$

$(9):$

$$\begin{array}{c|c} b & \vdash b \\ & \llcorner c \\ c & b \end{array} \qquad \vdash\!\!\!\begin{array}{l} a \\ \llcorner b \\ a \\ b \\ c \end{array} \qquad (11).$$

We can translate this formula thus: "If the proposition that b takes place or c does not is a sufficient condition for a, then b is by itself a sufficient condition for a".

$$\overset{8}{d \mid c} \qquad \vdash\!\!\!\begin{array}{l} a \\ \llcorner c \\ b \\ a \\ b \\ c \end{array}$$

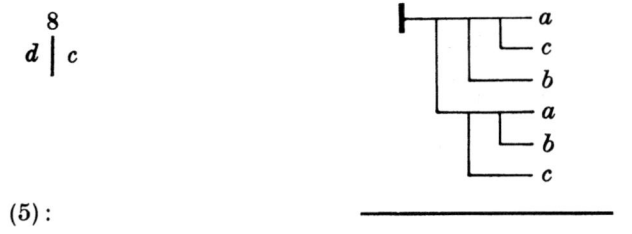

$(5):$

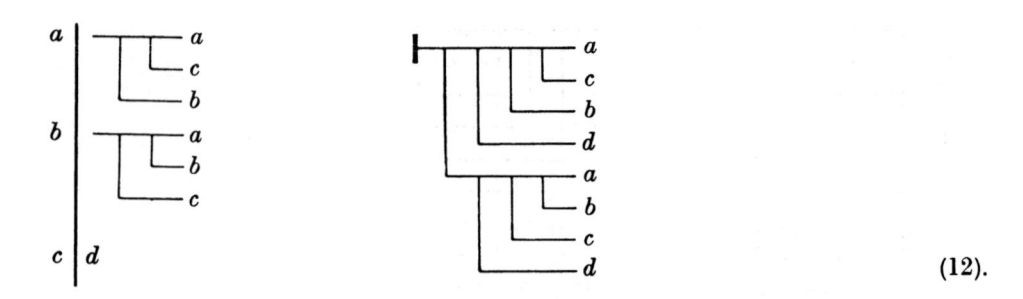

Propositions (12)–(17) and (22) show how, when there are several conditions, their order can be changed.

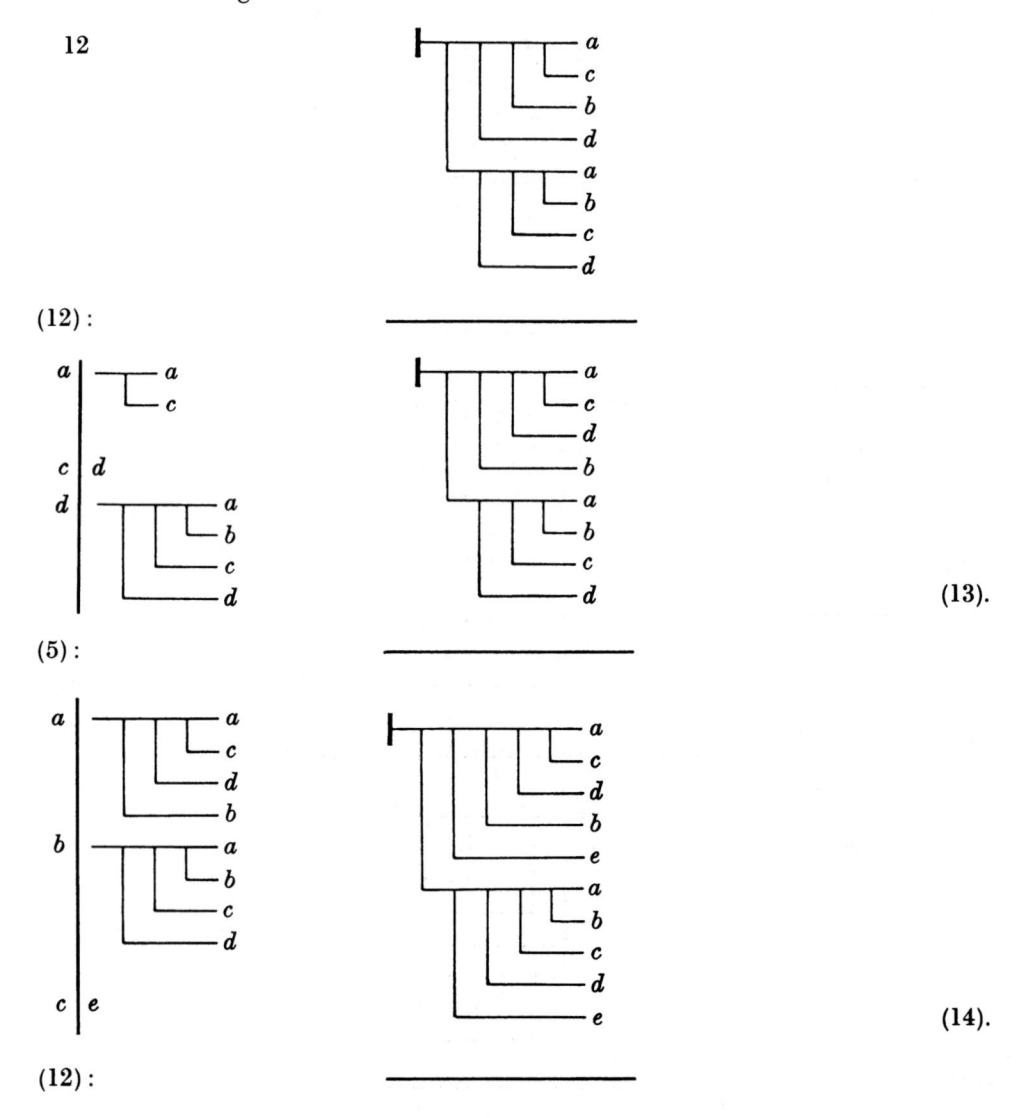

(12)

BEGRIFFSSCHRIFT

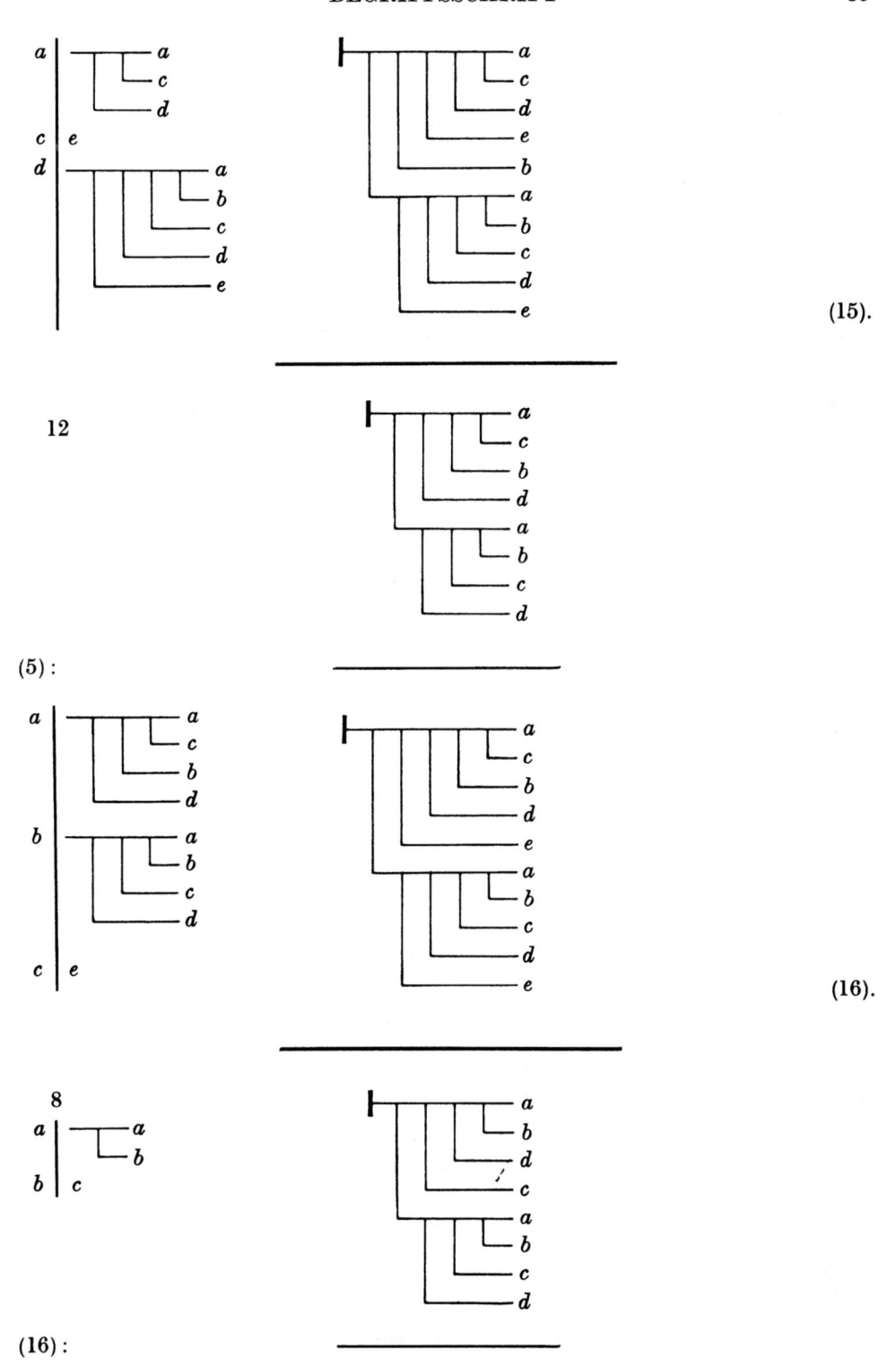

(15).

(5):

(16):

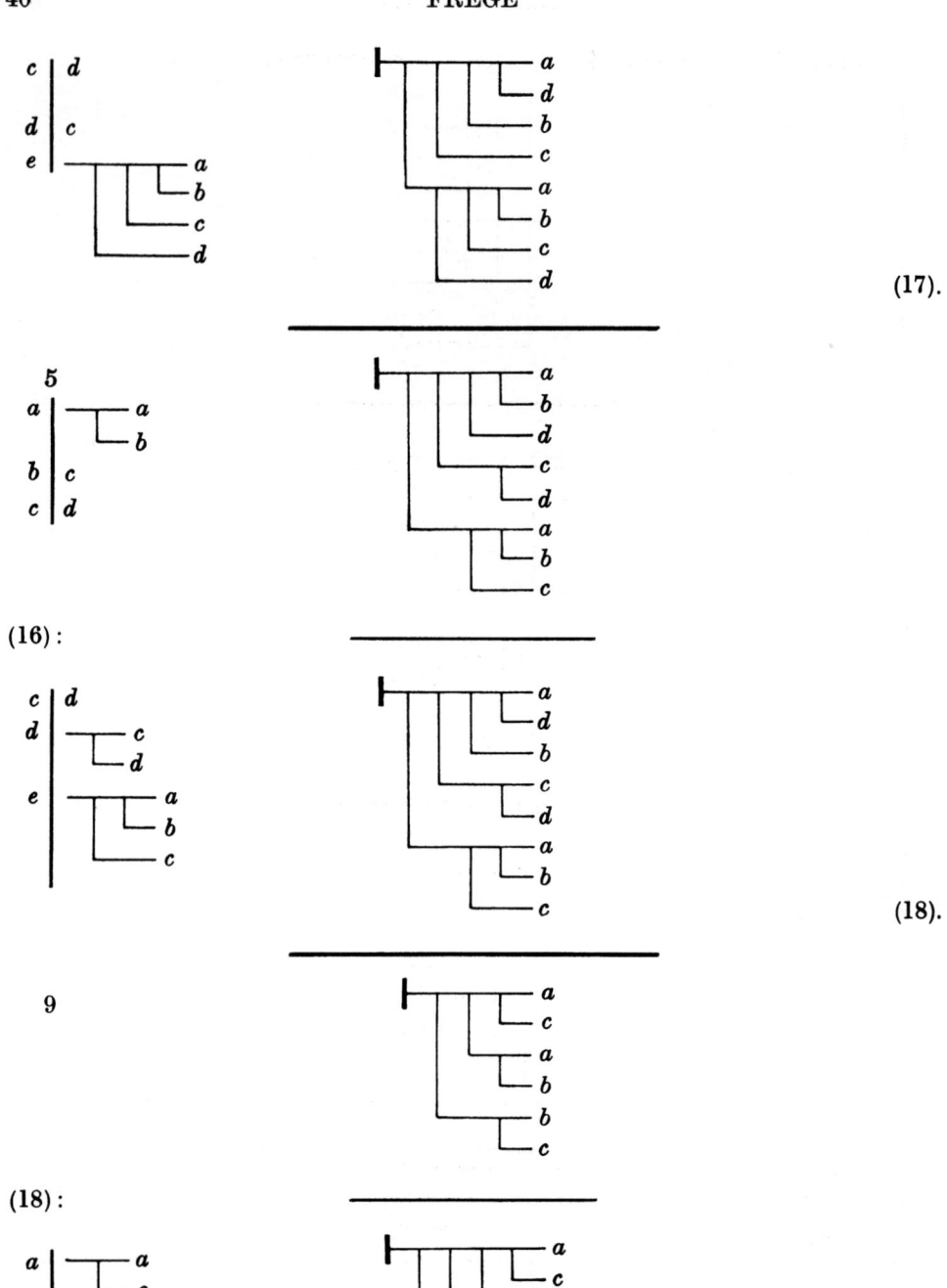

BEGRIFFSSCHRIFT

This proposition differs from (7) only in an unessential way.

19

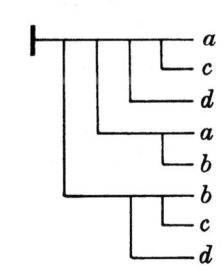

(18):

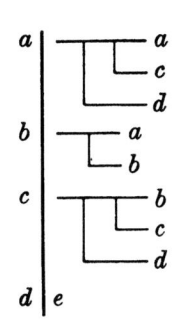

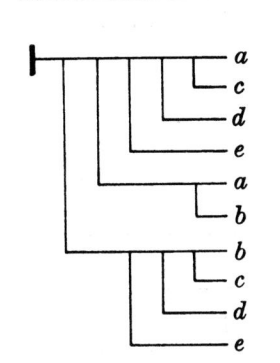

9

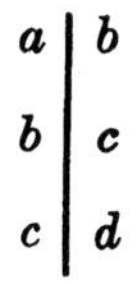

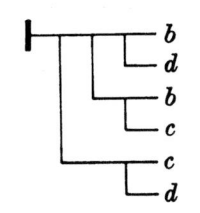

(19):

(20).

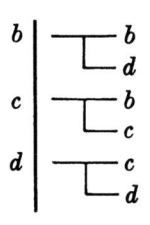

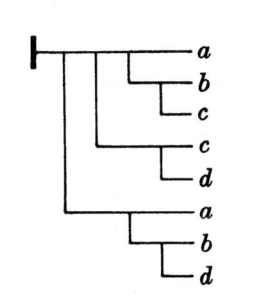

(21).

16

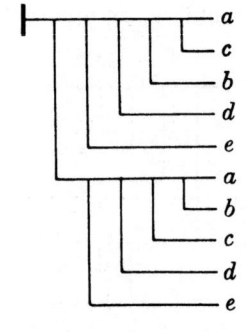

(5):

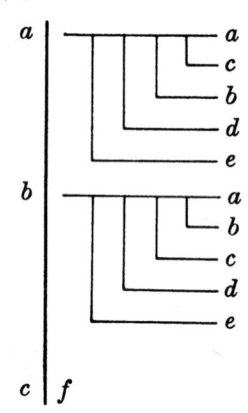

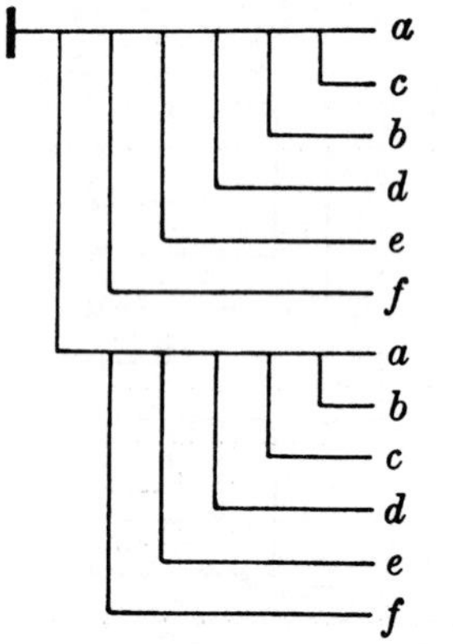

a	...
b	...
c	f

18

a	a, b
b	c
c	d
d	e

(22):

c	e
d	c
e	d, e
f	a, b, c, d

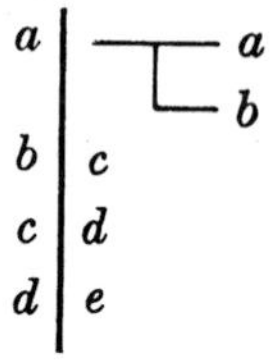

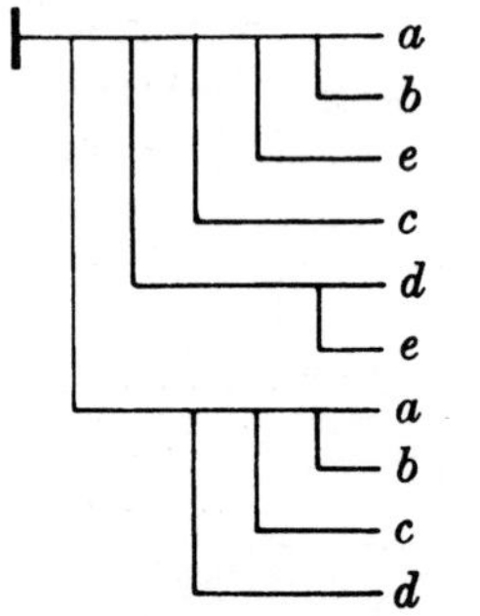

(22).

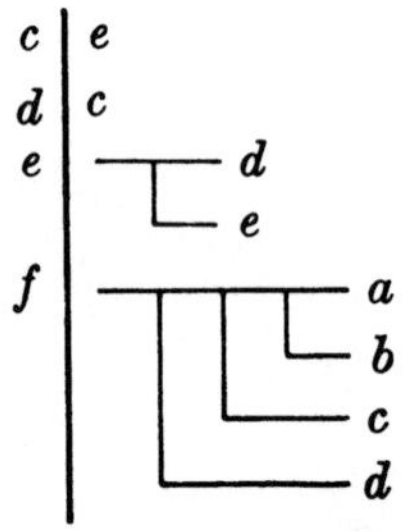

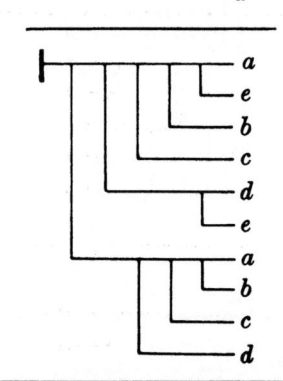

(23).

BEGRIFFSSCHRIFT 43

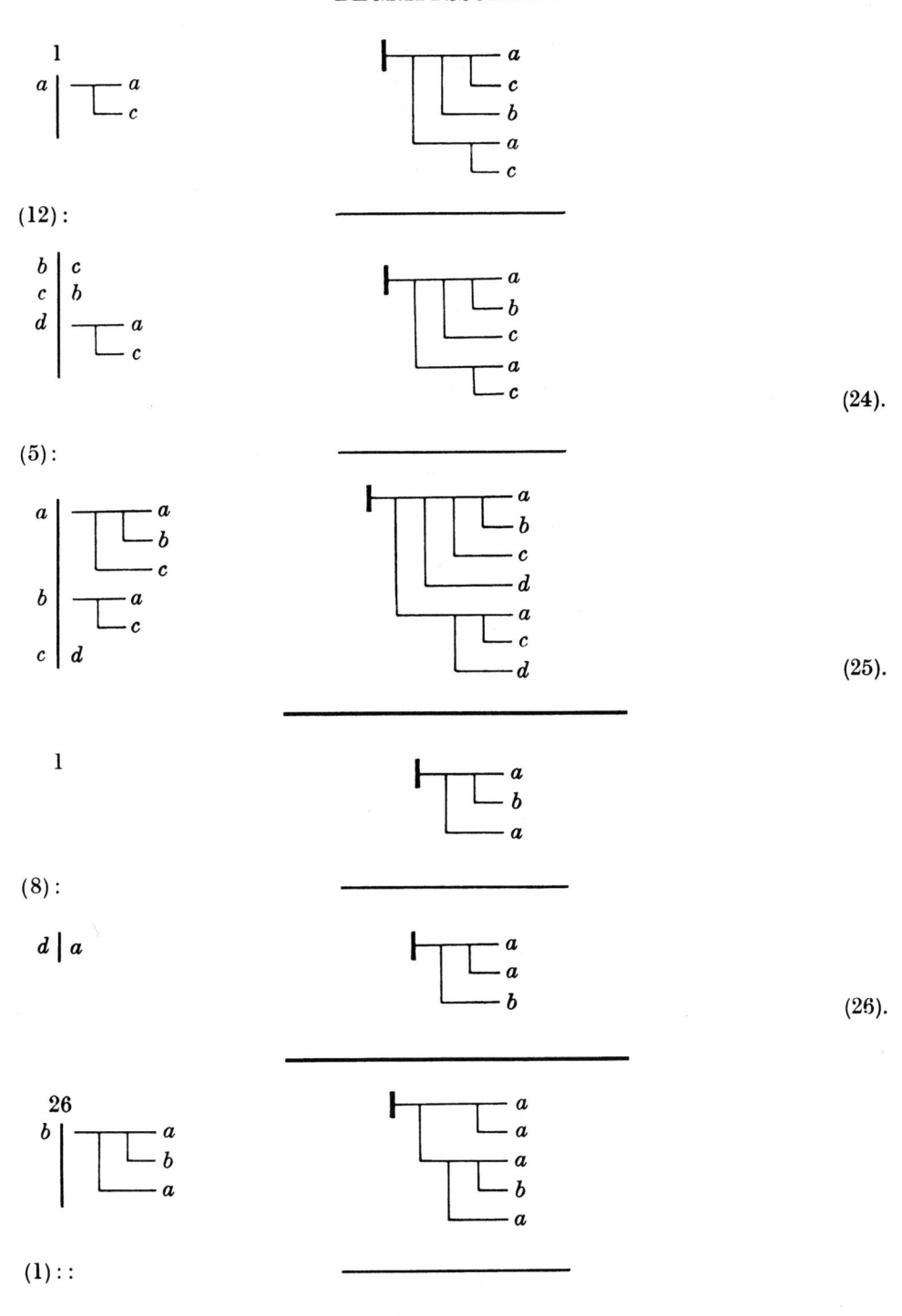

$$
\begin{array}{c|c}
 & 1 \\
a & \begin{array}{c} a \\ c \end{array}
\end{array}
$$

(12):

$$
\begin{array}{c|c}
b & c \\
c & b \\
d & \begin{array}{c} a \\ c \end{array}
\end{array}
$$

(5):

$$
\begin{array}{c|c}
a & \begin{array}{c} a \\ b \\ c \end{array} \\
b & \begin{array}{c} a \\ c \end{array} \\
c & d
\end{array}
$$

(8):

$$
\begin{array}{c|c}
 & 1 \\
d & a
\end{array}
$$

$$
\begin{array}{c|c}
 & 26 \\
b & \begin{array}{c} a \\ b \\ a \end{array}
\end{array}
$$

(1)::

(24).

(25).

(26).

(27).

FREGE

We cannot (at the same time) affirm a and deny a.

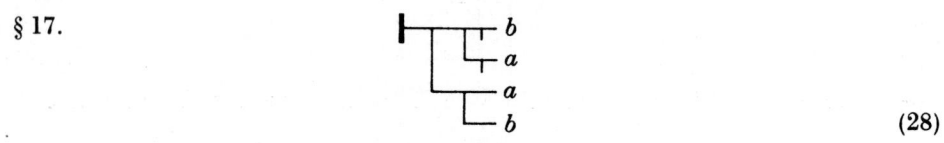

§ 17.

$$\vdash\!\!\begin{array}{l}\rule{0pt}{0pt}\!\!\top\!\!-b\\ \!\!\top\!\!-a\\ \!\!\top\!\!-a\\ \!\!\top\!\!-b\end{array} \qquad (28)$$

means: "The case in which $\top\!\!\begin{array}{l}-b\\ -a\end{array}$ is denied and $\top\!\!\begin{array}{l}-a\\ -b\end{array}$ is affirmed does not take place". The denial of $\top\!\!\begin{array}{l}-b\\ -a\end{array}$ means that $\top\!-a$ is affirmed and $\top\!-b$ is denied, that is, that a is denied and b is affirmed. This case is excluded by $\top\!\!\begin{array}{l}-a\\ -b\end{array}$. This judgment justifies the transition from *modus ponens* to *modus tollens*. For example, let b mean the proposition that the man M is alive, and a the proposition that M breathes. Then we have the judgment: "If from the circumstance that M is alive his breathing can be inferred, then from the circumstance that he does not breathe his death can be inferred".

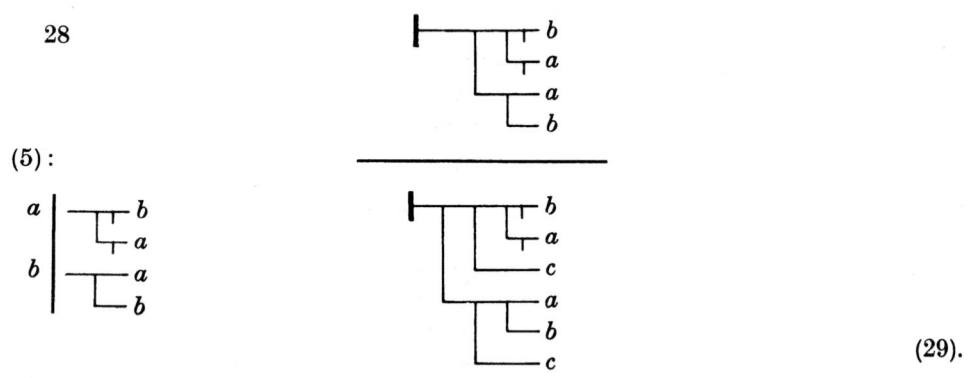

28

$$\vdash\!\!\begin{array}{l}\!\!\top\!\!-b\\ \!\!\top\!\!-a\\ \!\!\top\!\!-a\\ \!\!\top\!\!-b\end{array}$$

(5):

$$\begin{array}{c|c}a & \top\!\!\begin{array}{l}-b\\ -a\end{array}\\ b & \top\!\!\begin{array}{l}-a\\ -b\end{array}\end{array} \qquad \vdash\!\!\begin{array}{l}\!\!\top\!\!-b\\ \!\!\top\!\!-a\\ \!\!\top\!\!-c\\ \!\!\top\!\!-a\\ \!\!\top\!\!-b\\ \!\!-c\end{array} \qquad (29).$$

If b and c together form a sufficient condition for a, then from the affirmation of one condition, c, and that of the negation of a 〚that of〛 the negation of the other condition can be inferred.

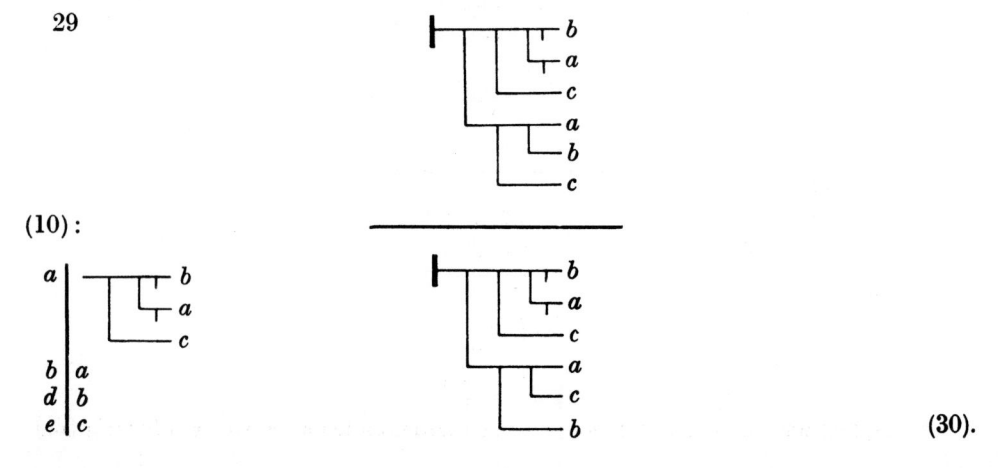

29

$$\vdash\!\!\begin{array}{l}\!\!\top\!\!-b\\ \!\!\top\!\!-a\\ \!\!-c\\ \!\!-a\\ \!\!-b\\ \!\!-c\end{array}$$

(10):

$$\begin{array}{c|c}a & \top\!\!\begin{array}{l}-b\\ -a\\ -c\end{array}\\ b & a\\ d & b\\ e & c\end{array} \qquad \vdash\!\!\begin{array}{l}\!\!\top\!\!-b\\ \!\!\top\!\!-a\\ \!\!-c\\ \!\!-a\\ \!\!-c\\ \!\!-b\end{array} \qquad (30).$$

§ 18.

$$\vdash\negthinspace\neg\negthinspace\neg a \atop \neg\neg a \qquad (31).$$

$\neg\neg a$ means the denial of the denial, hence the affirmation of a. Thus a cannot be denied and (at the same time) $\neg\neg a$ affirmed. *Duplex negatio affirmat.* The denial of the denial is affirmation.

$$\begin{array}{c} 31 \\ a \mid b \end{array} \qquad \vdash\negthinspace\begin{array}{l} b \\ \neg\neg b \end{array}$$

(7):

$$\begin{array}{c|l} a & b \\ b & \neg\neg b \\ c & \neg a \\ d & a \\ & b \end{array} \qquad \vdash\negthinspace\begin{array}{l} b \\ a \\ a \\ b \\ \neg\neg b \\ a \\ a \\ b \end{array} \qquad (32).$$

(28)::

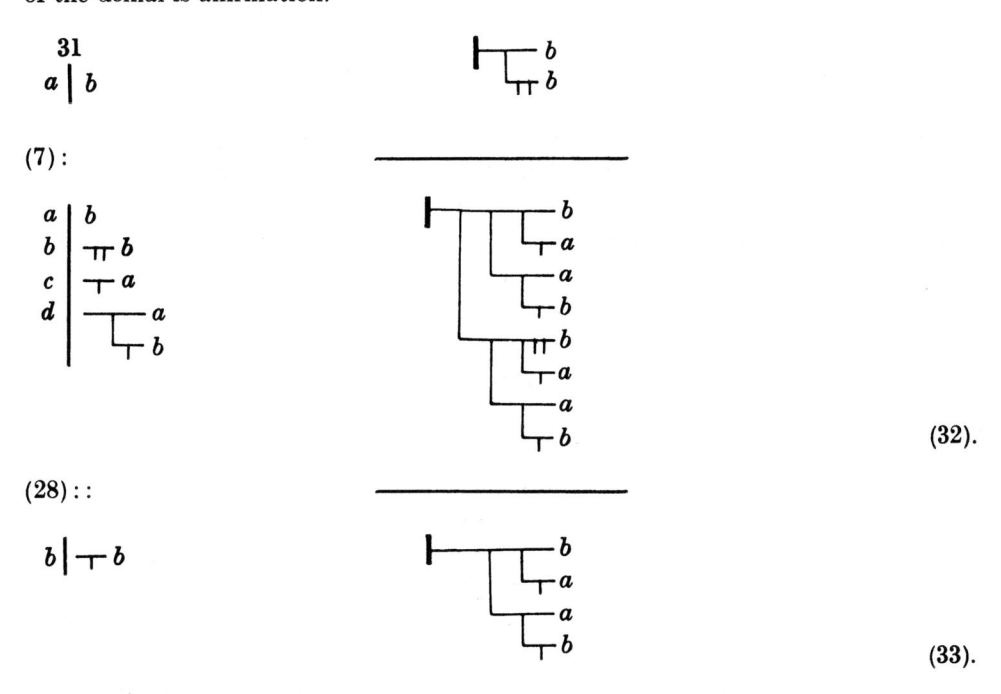

$$b \mid \neg b \qquad \vdash\negthinspace\begin{array}{l} b \\ a \\ a \\ b \end{array} \qquad (33).$$

If a or b takes place, then b or a takes place.

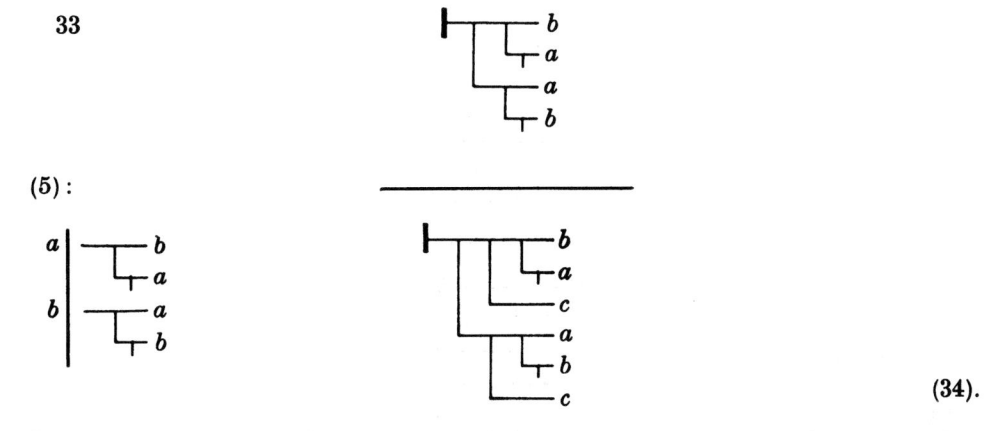

33

$$\vdash\negthinspace\begin{array}{l} b \\ a \\ a \\ b \end{array}$$

(5):

$$\begin{array}{c|l} a & b \\ & a \\ b & a \\ & b \end{array} \qquad \vdash\negthinspace\begin{array}{l} b \\ a \\ c \\ a \\ b \\ c \end{array} \qquad (34).$$

If as a consequence of the occurrence of the circumstance c, when the obstacle b is removed, a takes place, then from the circumstance that a does not take place while c occurs the occurrence of the obstacle b can be inferred.

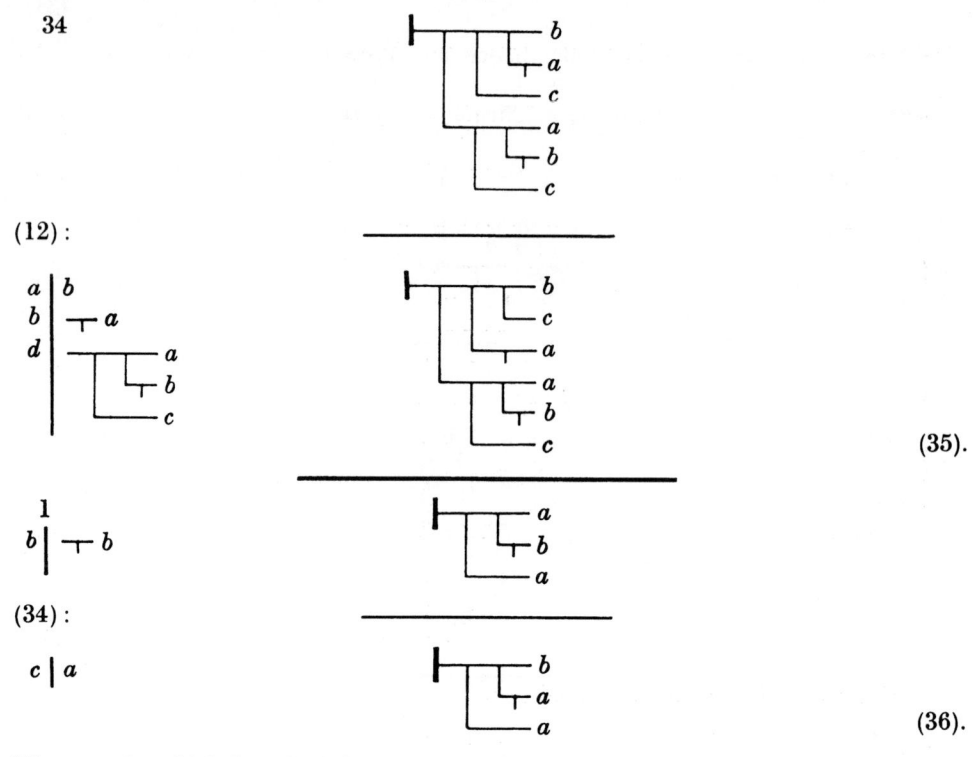

34

(12):

$$\dfrac{\begin{array}{c} a \mid b \\ b \mid {\displaystyle \mathop{\text{---}}_{\top}} a \\ d \end{array} \quad \begin{array}{c} a \\ b \\ c \end{array}}{\begin{array}{c} 1 \\ b \mid {\displaystyle \mathop{\text{---}}_{\top}} b \end{array}} \tag{35}.$$

(34):

$$c \mid a \tag{36}.$$

The case in which b is denied, $\dfrac{}{\top} a$ is affirmed, and a is affirmed does not occur. We can express this as follows: "If a occurs, then one of the two, a or b, takes place".

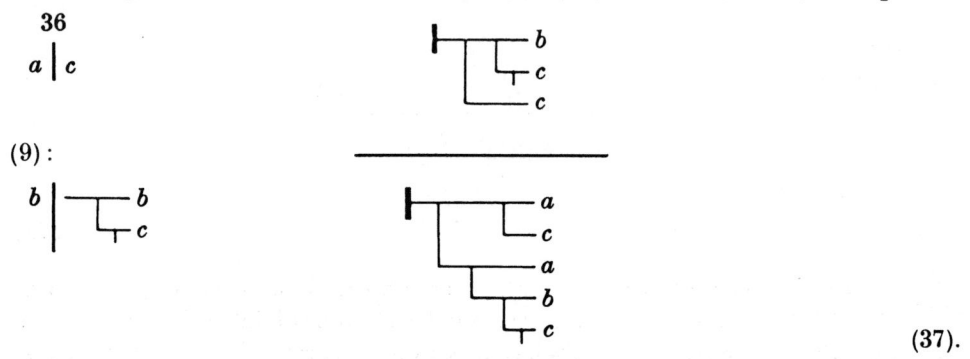

36
$a \mid c$

(9):

$$b \mid \dfrac{b}{c} \tag{37}.$$

If a is a necessary consequence of the occurrence of b or c, then a is a necessary consequence of c alone. For example, let b mean the circumstance that the first factor of a product P is 0, c the circumstance that the second factor of P is 0, and a the circumstance that the product P is 0. Then we have the judgment: "If the product P is 0 in case the first or the second factor is 0, then from the vanishing of the second factor the vanishing of the product can be inferred".

BEGRIFFSSCHRIFT

36

$$\vdash \!\!\begin{array}{l} \rule{0.4em}{0.4pt}\, b \\ \rule{0.4em}{0.4pt}\, a \\ \rule{0.4em}{0.4pt}\, a \end{array}$$

(8):

$$\begin{array}{c|c} a & b \\ b & \top a \\ d & a \end{array} \qquad \vdash \!\!\begin{array}{l} \rule{0.4em}{0.4pt}\, b \\ \rule{0.4em}{0.4pt}\, a \\ \rule{0.4em}{0.4pt}\, a \end{array} \qquad (38).$$

(2):

$$\begin{array}{c|c} a & b \\ b & a \\ c & \top a \end{array} \qquad \vdash \!\!\begin{array}{l} \rule{0.4em}{0.4pt}\, b \\ \rule{0.4em}{0.4pt}\, a \\ \rule{0.4em}{0.4pt}\, a \\ \rule{0.4em}{0.4pt}\, a \end{array} \qquad (39).$$

(35):

$$\begin{array}{c|c} a & b \\ b & a \\ c & \begin{array}{c} \rule{0.4em}{0.4pt}\, a \\ \rule{0.4em}{0.4pt}\, a \end{array} \end{array} \qquad \vdash \!\!\begin{array}{l} \rule{0.4em}{0.4pt}\, a \\ \rule{0.4em}{0.4pt}\, a \\ \rule{0.4em}{0.4pt}\, a \\ \rule{0.4em}{0.4pt}\, b \end{array} \qquad (40).$$

§ 19.

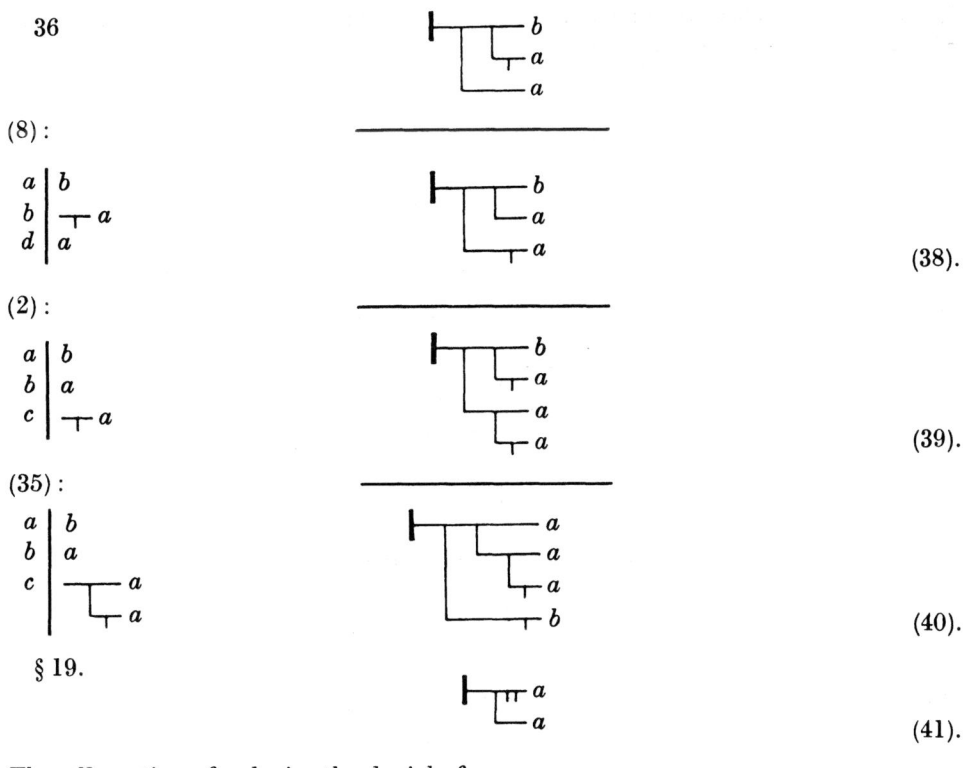

$$\vdash \!\!\begin{array}{l} \rule{0.4em}{0.4pt}\, a \\ \rule{0.4em}{0.4pt}\, a \end{array} \qquad (41).$$

The affirmation of a denies the denial of a.

27

$$\vdash \!\!\begin{array}{l} \rule{0.4em}{0.4pt}\, a \\ \rule{0.4em}{0.4pt}\, a \end{array}$$

(41):

$$a \;\Big|\; \begin{array}{c} \rule{0.4em}{0.4pt}\, a \\ \rule{0.4em}{0.4pt}\, a \end{array} \qquad \vdash \!\!\begin{array}{l} \rule{0.4em}{0.4pt}\, a \\ \rule{0.4em}{0.4pt}\, a \end{array} \qquad (42).$$

(40):

$$b \;\Big|\; \begin{array}{c} \rule{0.4em}{0.4pt}\, a \\ \rule{0.4em}{0.4pt}\, a \end{array} \qquad \vdash \!\!\begin{array}{l} \rule{0.4em}{0.4pt}\, a \\ \rule{0.4em}{0.4pt}\, a \\ \rule{0.4em}{0.4pt}\, a \end{array} \qquad (43).$$

If there is a choice only between a and a, then a takes place. For example, we have to distinguish two cases that between them exhaust all possibilities. In following the first, we arrive at the result that a takes place; the same result holds when we follow the second. Then the proposition a holds.

43

$$\vdash \!\!\begin{array}{l} \rule{0.4em}{0.4pt}\, a \\ \rule{0.4em}{0.4pt}\, a \\ \rule{0.4em}{0.4pt}\, a \end{array}$$

(21):

(5):

(33)::

If a holds when c occurs as well as when c does not occur, then a holds. Another way of expressing it is: "If a or c occurs and if the occurrence of c has a as a necessary consequence, then a takes place".

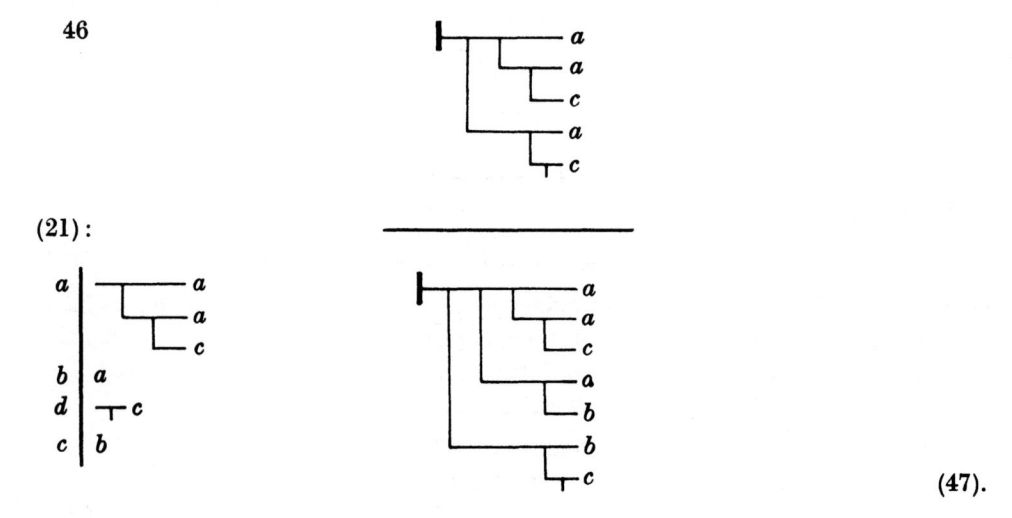

46

(21):

We can express this proposition thus: "If c, as well as b, is a sufficient condition for a and if b or c takes place, then the proposition a holds". This judgment is used when

BEGRIFFSSCHRIFT

two cases are to be distinguished in a proof. When more cases occur, we can always reduce them to two by taking one of the cases as the first and the totality of the others as the second. The latter can in turn be broken down into two cases, and this can be continued so long as further decomposition is possible.

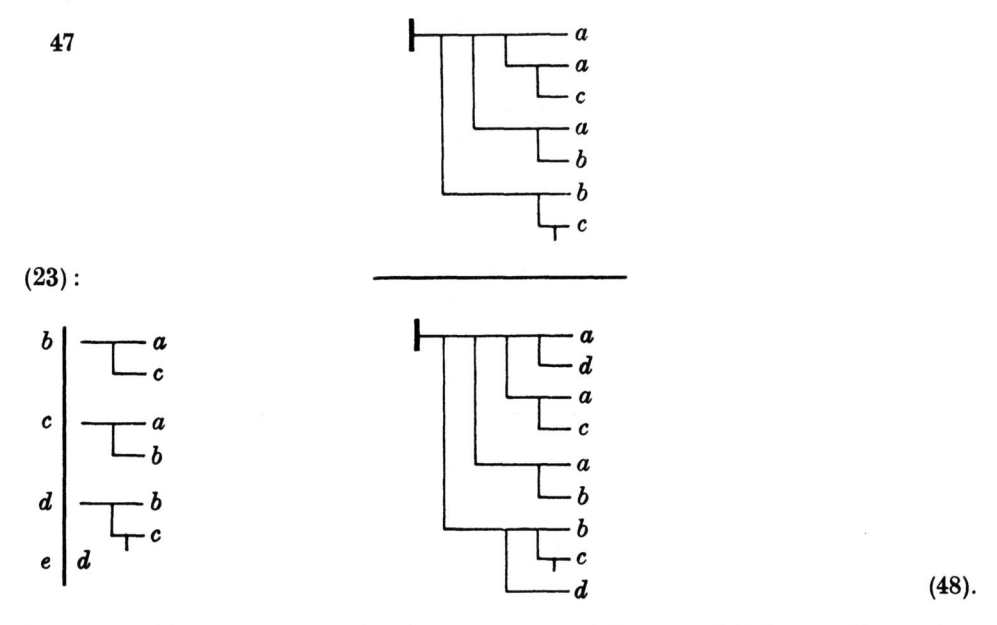

47

(23):

(48).

If d is a sufficient condition for the occurrence of b or c and if b, as well as c, is a sufficient condition for a, then d is a sufficient condition for a. An example of an application is furnished by the derivation of formula (101).

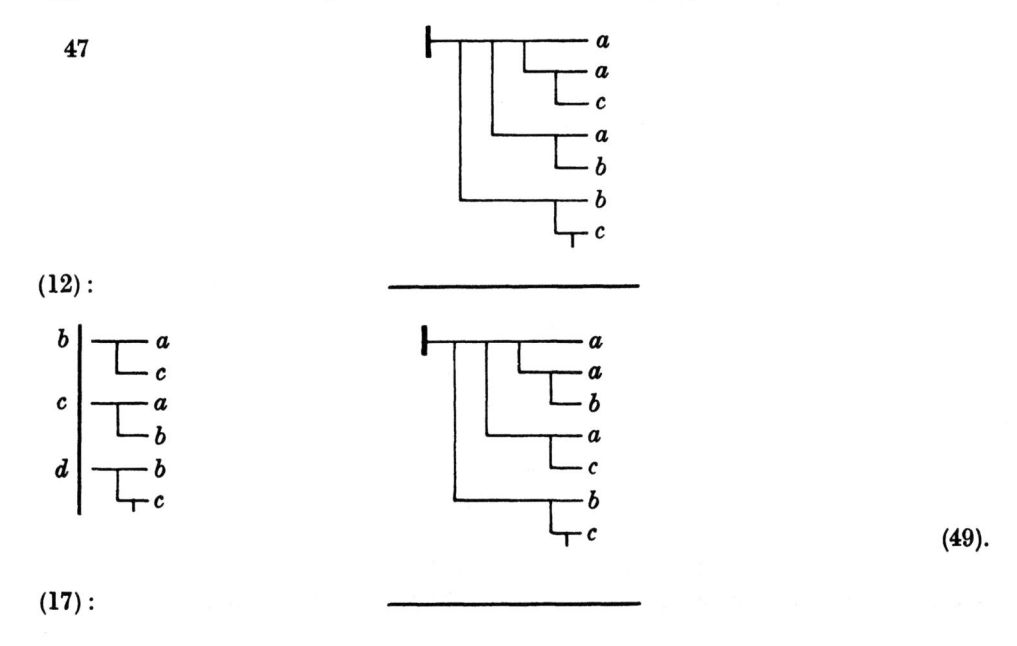

47

(12):

(49).

(17):

(18):

$$\begin{array}{c|c} b & a \\ & b \\ c & a \\ & c \\ d & b \\ & c \end{array}$$

$$\vdash f(a,b,c,a,b,a,c) \quad (50).$$

$$\begin{array}{c|c} a & a \\ & b \\ & c \\ b & a \\ & b \\ c & a \\ & c \end{array}$$

$$\vdash f(a,b,c,d,a,b,a,c,d) \quad (51).$$

§ 20.

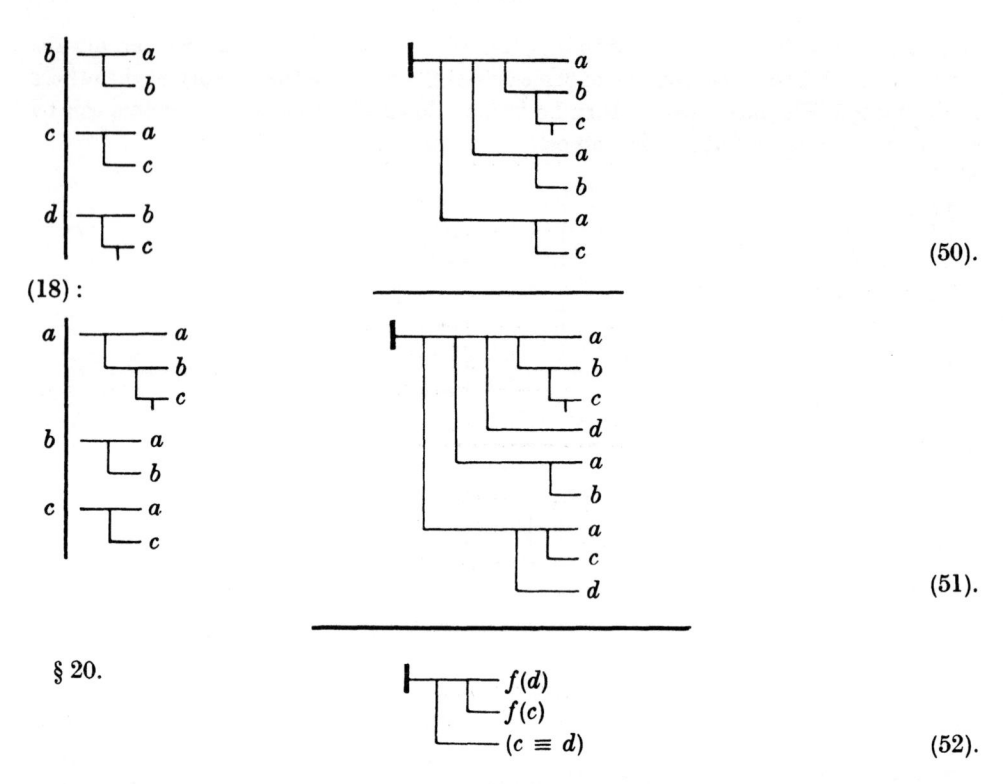

$$\vdash \begin{array}{c} f(d) \\ f(c) \\ (c \equiv d) \end{array} \quad (52).$$

The case in which the content of c is identical with the content of d and in which $f(c)$ is affirmed and $f(d)$ is denied does not take place. This proposition means that, if $c \equiv d$, we could everywhere put d for c. In $f(c)$, c can also occur in other than the argument places. Hence c may still be contained in $f(d)$.

52

$$\vdash \begin{array}{c} f(d) \\ f(c) \\ (c \equiv d) \end{array}$$

(8):

$$\begin{array}{c|c} a & f(d) \\ b & f(c) \\ d & (c \equiv d) \end{array}$$

$$\vdash \begin{array}{c} f(d) \\ (c \equiv d) \\ f(c) \end{array} \quad (53).$$

§ 21.

$$\vdash (c \equiv c) \quad (54).$$

The content of c is identical with the content of c.

54

$$\vdash (c \equiv c)$$

(53):

$$f(A) \,\big|\, (A \equiv c) \qquad\qquad \vdash \begin{array}{l}(d \equiv c) \\ (c \equiv d)\end{array} \qquad\qquad (55).$$

(9):

$$\begin{array}{l|l} b & (d \equiv c) \\ c & (c \equiv d) \\ a & \begin{array}{l} f(c) \\ f(d) \end{array} \end{array} \qquad \vdash \begin{array}{l} f(c) \\ f(d) \\ (c \equiv d) \\ f(c) \\ f(d) \\ (d \equiv c) \end{array} \qquad (56).$$

(52)::

$$\begin{array}{l|l} d & c \\ c & d \end{array} \qquad\qquad \vdash \begin{array}{l} f(c) \\ f(d) \\ (c \equiv d) \end{array} \qquad\qquad (57).$$

§ 22.

$$\vdash \begin{array}{l} f(c) \\ \rotatebox{0}{$\smile^{\mathfrak{a}}$} f(\mathfrak{a}) \end{array} \qquad\qquad (58).$$

$\smile^{\mathfrak{a}}\!\!-\! f(\mathfrak{a})$ means that $f(\mathfrak{a})$ takes place, whatever we may understand by $\mathfrak{a}$. If therefore $\smile^{\mathfrak{a}}\!\!-\! f(\mathfrak{a})$ is affirmed, $f(c)$ cannot be denied. This is what our proposition expresses.

Here $\mathfrak{a}$ can occur only in the argument places of f, since in the judgment this function also occurs outside the scope of $\mathfrak{a}$.

$$\begin{array}{c} 58 \\ f(A) \,\big|\, \begin{array}{l} f(A) \\ g(A) \end{array} \\ c \;\; b \end{array} \qquad\qquad \vdash \begin{array}{l} f(b) \\ g(b) \\ \smile^{\mathfrak{a}}\; \begin{array}{l} f(\mathfrak{a}) \\ g(\mathfrak{a}) \end{array} \end{array}$$

(30):

$$\begin{array}{l|l} a & f(b) \\ c & g(b) \\ b & \smile^{\mathfrak{a}}\; \begin{array}{l} f(\mathfrak{a}) \\ g(\mathfrak{a}) \end{array} \end{array} \qquad \vdash \begin{array}{l} \smile^{\mathfrak{a}} \begin{array}{l} f(\mathfrak{a}) \\ g(\mathfrak{a}) \end{array} \\ f(b) \\ g(b) \end{array} \qquad (59).$$

Example. Let b mean an ostrich, that is, an individual animal belonging to the species, let $g(A)$ mean "A is a bird", and let $f(A)$ mean "A can fly". Then we have the judgment "If this ostrich is a bird and cannot fly, then it can be inferred from this that some[17] birds cannot fly".

[17] See footnote 16.

We see how this judgment replaces one mode of inference, namely, Felapton or Fesapo, between which we do not distinguish here since no subject has been singled out.

(60).

(61).

(62).

This judgment replaces the mode of inference Barbara when the minor premiss, $g(x)$, has a particular content.

BEGRIFFSSCHRIFT

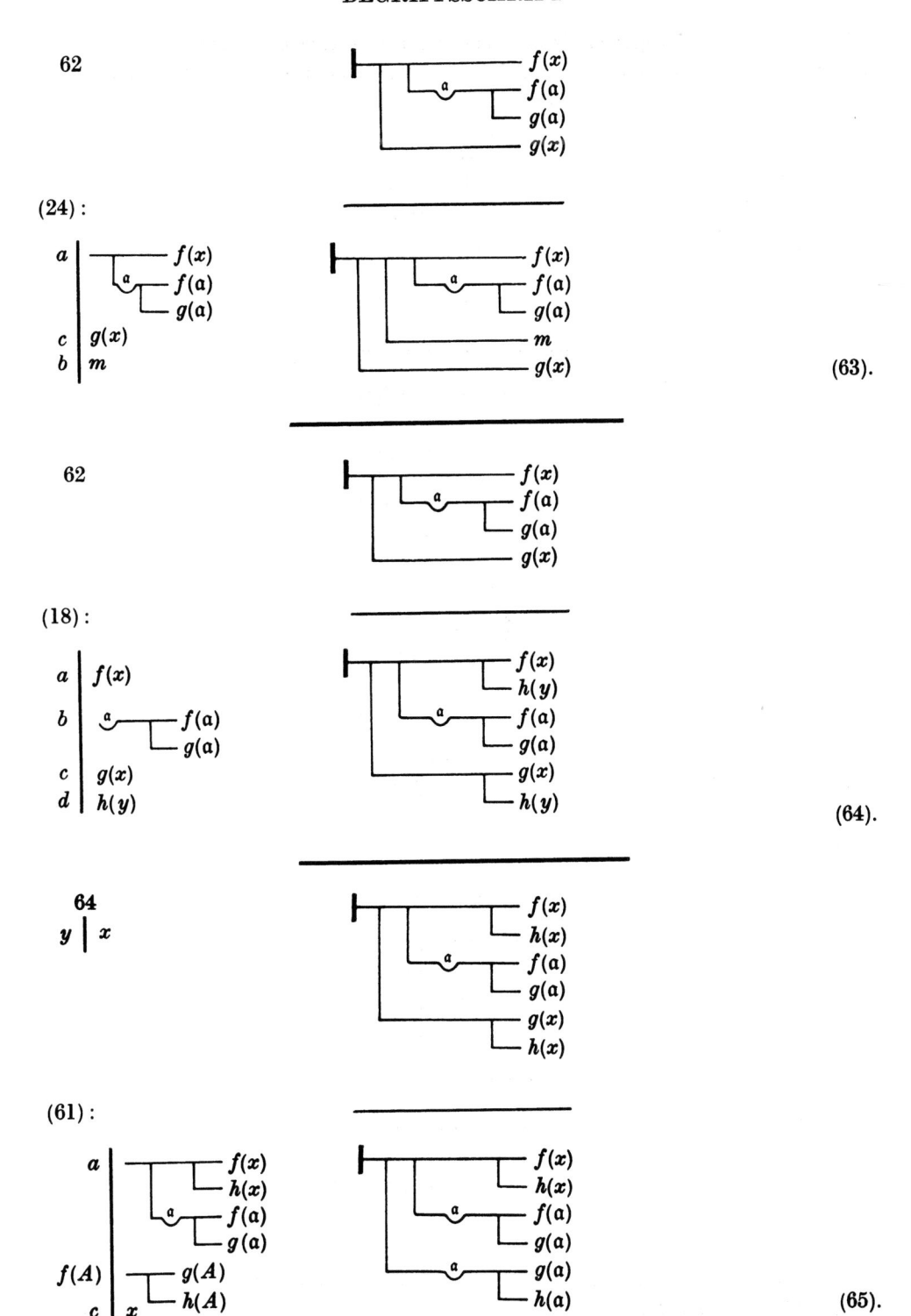

53

62

$$\vdash\!\!\!\begin{array}{l} \rule{1em}{0.4pt}\ f(x) \\ \rule{1em}{0.4pt}\ f(\mathfrak{a}) \\ \rule{1em}{0.4pt}\ g(\mathfrak{a}) \\ \rule{1em}{0.4pt}\ g(x) \end{array}$$

(24):

$$\begin{array}{c|l} a & \begin{array}{l} f(x) \\ f(\mathfrak{a}) \\ g(\mathfrak{a}) \end{array} \\ c & g(x) \\ b & m \end{array} \qquad \vdash\!\!\!\begin{array}{l} f(x) \\ f(\mathfrak{a}) \\ g(\mathfrak{a}) \\ m \\ g(x) \end{array} \qquad (63).$$

62

$$\vdash\!\!\!\begin{array}{l} f(x) \\ f(\mathfrak{a}) \\ g(\mathfrak{a}) \\ g(x) \end{array}$$

(18):

$$\begin{array}{c|l} a & f(x) \\ b & \begin{array}{l} f(\mathfrak{a}) \\ g(\mathfrak{a}) \end{array} \\ c & g(x) \\ d & h(y) \end{array} \qquad \vdash\!\!\!\begin{array}{l} f(x) \\ h(y) \\ f(\mathfrak{a}) \\ g(\mathfrak{a}) \\ g(x) \\ h(y) \end{array} \qquad (64).$$

$$\begin{array}{c} 64 \\ y \mid x \end{array} \qquad \vdash\!\!\!\begin{array}{l} f(x) \\ h(x) \\ f(\mathfrak{a}) \\ g(\mathfrak{a}) \\ g(x) \\ h(x) \end{array}$$

(61):

$$\begin{array}{c|l} a & \begin{array}{l} f(x) \\ h(x) \\ f(\mathfrak{a}) \\ g(\mathfrak{a}) \end{array} \\ f(A) & g(A) \\ c & x \quad h(A) \end{array} \qquad \vdash\!\!\!\begin{array}{l} f(x) \\ h(x) \\ f(\mathfrak{a}) \\ g(\mathfrak{a}) \\ g(\mathfrak{a}) \\ h(\mathfrak{a}) \end{array} \qquad (65).$$

Here $\mathfrak{a}$ occurs in two scopes, but this does not indicate any particular relation between them. In one of these scopes we could also write, say, $\mathfrak{e}$ instead of $\mathfrak{a}$. This judgment replaces the mode of inference Barbara when the minor premiss

$$\vdash\!\!-\!\!\cup^{\mathfrak{a}}\!\!-\!\!\begin{array}{l} g(\mathfrak{a}) \\ h(\mathfrak{a}) \end{array}$$

has a general content. The reader who has familiarized himself with the way derivations are carried out in the ideography will be in a position to derive also the judgments that answer to the other modes of inference. These should suffice as examples here.

$$65 \qquad \vdash\!\!-\!\!\begin{array}{l} f(x) \\ h(x) \\ \cup^{\mathfrak{a}}\!\!-\!\!\begin{array}{l} f(\mathfrak{a}) \\ g(\mathfrak{a}) \end{array} \\ \cup^{\mathfrak{a}}\!\!-\!\!\begin{array}{l} g(\mathfrak{a}) \\ h(\mathfrak{a}) \end{array} \end{array}$$

$$(8): \quad \begin{array}{c|l} \mathfrak{a} & f(x),\ h(x) \\ \mathfrak{b} & \cup^{\mathfrak{a}}\!-\! f(\mathfrak{a}),\ g(\mathfrak{a}) \\ \mathfrak{d} & \cup^{\mathfrak{a}}\!-\! g(\mathfrak{a}),\ h(\mathfrak{a}) \end{array} \qquad \vdash\!\!-\!\!\begin{array}{l} f(x) \\ h(x) \\ \cup^{\mathfrak{a}}\!\!-\!\!\begin{array}{l} g(\mathfrak{a}) \\ h(\mathfrak{a}) \end{array} \\ \cup^{\mathfrak{a}}\!\!-\!\!\begin{array}{l} f(\mathfrak{a}) \\ g(\mathfrak{a}) \end{array} \end{array} \qquad (66).$$

$$58 \qquad \vdash\!\!-\!\!\begin{array}{l} f(c) \\ \cup^{\mathfrak{a}}\!\!-\! f(\mathfrak{a}) \end{array}$$

$$(7): \quad \begin{array}{c|l} \mathfrak{a} & f(c) \\ \mathfrak{b} & \cup^{\mathfrak{a}}\!-\! f(\mathfrak{a}) \\ \mathfrak{c} & b \\ \mathfrak{d} & [(\cup^{\mathfrak{a}}\!-\! f(\mathfrak{a})) \equiv b] \end{array} \qquad \vdash\!\!-\!\!\begin{array}{l} f(c) \\ b \\ [(\cup^{\mathfrak{a}}\!-\! f(\mathfrak{a})) \equiv b] \\ \cup^{\mathfrak{a}}\!-\!\begin{array}{l} f(\mathfrak{a}) \\ b \\ [(\cup^{\mathfrak{a}}\!-\! f(\mathfrak{a})) \equiv b] \end{array} \end{array} \qquad (67).$$

$$(57):: \quad \begin{array}{c|l} f(A) & A \\ c & \cup^{\mathfrak{a}}\!-\! f(\mathfrak{a}) \\ d & b \end{array} \qquad \vdash\!\!-\!\!\begin{array}{l} f(c) \\ b \\ [(\cup^{\mathfrak{a}}\!-\! f(\mathfrak{a})) \equiv b] \end{array} \qquad (68).$$

III. SOME TOPICS FROM A GENERAL THEORY OF SEQUENCES

§ 23. The derivations that follow are intended to give a general idea of the way in which our ideography is handled, even if they are perhaps not sufficient to demonstrate its full utility. This utility would become clear only when more involved propositions are considered. Through the present example, moreover, we see how pure thought, irrespective of any content given by the senses or even by an intuition a priori, can, solely from the content that results from its own constitution, bring forth judgments that at first sight appear to be possible only on the basis of some intuition. This can be compared with condensation, through which it is possible to transform the air that to a child's consciousness appears as nothing into a visible fluid that forms drops. The propositions about sequences developed in what follows far surpass in generality all those that can be derived from any intuition of sequences. If, therefore, one were to consider it more appropriate to use an intuitive idea of sequence as a basis, he should not forget that the propositions thus obtained, which might perhaps have the same wording as those given here, would still state far less than these, since they would hold only in the domain of precisely that intuition upon which they were based.

§ 24.
$$\Vdash \left[\begin{array}{c} \smile^{\mathfrak{b}} \smile^{\mathfrak{a}} F(\mathfrak{a}) \\ f(\mathfrak{b}, \mathfrak{a}) \\ F(\mathfrak{b}) \end{array} \right] \equiv \underset{\alpha}{\overset{\delta}{\big|}} \left(\begin{array}{c} F(\alpha) \\ f(\delta, \alpha) \end{array} \right) \qquad (69).$$

This proposition differs from the judgments considered up to now in that it contains signs that have not been defined before; it itself gives the definition. It does not say "The right side of the equation has the same content as the left", but "It is to have the same content". Hence this proposition is not a judgment, and consequently *not a synthetic judgment* either, to use the Kantian expression. I point this out because Kant considers all judgments of mathematics to be synthetic. If now (69) were a synthetic judgment, so would be the propositions derived from it. But we can do without the notation introduced by this proposition and hence without the proposition itself as its definition; nothing follows from the proposition that could not also be inferred without it. Our sole purpose in introducing such definitions is to bring about an extrinsic simplification by stipulating an abbreviation. They serve besides to emphasize a particular combination of signs in the multitude of possible ones, so that our faculty of representation can get a firmer grasp of it. Now, even though the simplification mentioned is hardly noticeable in the case of the small number of judgments cited here, I nevertheless included this formula for the sake of the example.

Although originally (69) is not a judgment, it is immediately transformed into one; for, once the meaning of the new signs is specified, it must remain fixed, and therefore formula (69) also holds as a judgment, but as an analytic one, since it only makes apparent again what was put into the new signs. This dual character of the formula is indicated by the use of a double judgment stroke. So far as the derivations that follow are concerned, (69) can therefore be treated like an ordinary judgment.

Lower-case Greek letters, which occur here for the first time, do not represent an independent content, as do German and Latin ones. The only thing we have to observe is whether they are identical or different; hence we can put arbitrary lower-case Greek letters for α and δ, provided only that places previously occupied by identical letters are again occupied by identical ones and that different letters are not replaced by identical ones. *Whether Greek letters are identical or different, however, is of significance only within the formula for which they were especially introduced,* as they were here for

$$\underset{\alpha}{\overset{\delta}{\bigm|}} \Bigl(\begin{matrix} F(\alpha) \\ f(\delta, \alpha) \end{matrix} \Bigr).$$

Their purpose is to enable us to reconstruct unambiguously at any time from the abbreviated form

$$\underset{\alpha}{\overset{\delta}{\bigm|}} \Bigl(\begin{matrix} F(\alpha) \\ f(\delta, \alpha) \end{matrix} \Bigr)$$

the full one,

For example,

$$\overset{\mathfrak{b}}{\frown}\overset{\mathfrak{a}}{\frown}\begin{matrix} F(\mathfrak{a}) \\ f(\mathfrak{b}, \mathfrak{a}) \\ F(\mathfrak{b}). \end{matrix}$$

$$\underset{\delta}{\overset{\alpha}{\bigm|}} \Bigl(\begin{matrix} F(\delta) \\ f(\delta, \alpha) \end{matrix} \Bigr)$$

means the expression

$$\overset{\mathfrak{b}}{\frown}\overset{\mathfrak{a}}{\frown}\begin{matrix} F(\mathfrak{a}) \\ f(\mathfrak{a}, \mathfrak{b}) \\ F(\mathfrak{b}), \end{matrix}$$

whereas

$$\underset{\delta}{\overset{\alpha}{\bigm|}} \Bigl(\begin{matrix} F(\alpha) \\ f(\delta, \alpha) \end{matrix} \Bigr)$$

has no meaning. We see that the complete expression, no matter how involved the functions F and f may be, can always be retrieved with certainty, except for the arbitrary choice of German letters.

$$\vdash\!\!-\!\!-f(\Gamma, \Delta)$$

can be rendered by "Δ is a result of an application of the procedure f to Γ", by "Γ is the object of an application of the procedure f, with the result Δ", by "Δ bears the relation f to Γ", or by "Γ bears the converse relation of f to Δ"; these expressions are to be taken as equivalent.

$$\underset{\alpha}{\overset{\delta}{\bigm|}} \Bigl(\begin{matrix} F(\alpha) \\ f(\delta, \alpha) \end{matrix} \Bigr)$$

can be translated by "the circumstance that property F is hereditary in the f-sequence [sich in der f-Reihe vererbt]". Perhaps the following example can make this expression acceptable. Let $\Lambda(M, N)$ mean the circumstance that N is a child of M, and $\Sigma(P)$ the circumstance that P is a human being. Then

$$\underset{\alpha}{\overset{\delta}{\,|\,}}\!\!\left(\begin{array}{l}\Sigma(\alpha)\\ \Lambda(\delta,\alpha)\end{array}\right. \quad \text{or} \quad \overset{\mathfrak{b}}{\smile}\overset{\mathfrak{a}}{\top}\!\!\begin{array}{l}\Sigma(\mathfrak{a})\\ \Lambda(\mathfrak{b},\mathfrak{a})\\ \Sigma(\mathfrak{b})\end{array}$$

is the circumstance that every child of a human being is in turn a human being, or that the property of being human is hereditary.[18] We see, incidentally, that it can become difficult and even impossible to give a rendering in words if very involved functions take the places of F and f. Proposition (69) could be expressed in words as follows:

If from the proposition that $\mathfrak{b}$ has property F it can be inferred generally, whatever $\mathfrak{b}$ may be, that every result of an application of the procedure f to $\mathfrak{b}$ has property F, then I say: "Property F is hereditary in the f-sequence".

§ 25.

69

$$\vdash\left[\left[\overset{\mathfrak{b}}{\smile}\overset{\mathfrak{a}}{\top}\!\!\begin{array}{l}F(\mathfrak{a})\\ f(\mathfrak{b},\mathfrak{a})\\ F(\mathfrak{b})\end{array}\right]\equiv\underset{\alpha}{\overset{\delta}{\,|\,}}\!\!\left(\begin{array}{l}F(\alpha)\\ f(\delta,\alpha)\end{array}\right.\right]$$

(68):

$$f(\Gamma)\ \bigg|\ \begin{array}{l}\mathfrak{a}\\ \mathfrak{b}\\ \mathfrak{b}\\ c\\ \end{array}\ \overset{\mathfrak{a}}{\smile}\top\!\!\begin{array}{l}F(\mathfrak{a})\\ f(\Gamma,\mathfrak{a})\\ F(\Gamma)\end{array}\ \underset{\alpha}{\overset{\delta}{\,|\,}}\!\!\left(\begin{array}{l}F(\alpha)\\ f(\delta,\alpha)\end{array}\right.$$

$$\vdash\ \overset{\mathfrak{a}}{\top}\!\!\begin{array}{l}F(\mathfrak{a})\\ f(x,\mathfrak{a})\\ F(x)\end{array}\ \underset{\alpha}{\overset{\delta}{\,|\,}}\!\!\left(\begin{array}{l}F(\alpha)\\ f(\delta,\alpha)\end{array}\right. \qquad (70).$$

(19):

$$\begin{array}{l}\mathfrak{b}\\ c\\ d\\ a\end{array}\ \bigg|\ \begin{array}{l}\overset{\mathfrak{a}}{\smile}\top\!\!\begin{array}{l}F(\mathfrak{a})\\ f(x,\mathfrak{a})\end{array}\\ F(x)\\ \underset{\alpha}{\overset{\delta}{\,|\,}}\!\!\left(\begin{array}{l}F(\alpha)\\ f(\delta,\alpha)\end{array}\right.\\ \top\!\!\begin{array}{l}F(y)\\ f(x,y)\end{array}\end{array} \qquad \vdash\ \begin{array}{l}F(y)\\ f(x,y)\\ F(x)\\ \underset{\alpha}{\overset{\delta}{\,|\,}}\!\!\left(\begin{array}{l}F(\alpha)\\ f(\delta,\alpha)\end{array}\right.\\ F(y)\\ f(x,y)\\ \overset{\mathfrak{a}}{\smile}\top\!\!\begin{array}{l}F(\mathfrak{a})\\ f(x,\mathfrak{a})\end{array}\end{array} \qquad (71).$$

(58)::

---

[18] [In the German text the formulas contain two misprints: at the extreme left "δ" and the "α" below it are interchanged, and, instead of "$\Lambda(\mathfrak{b},\mathfrak{a})$", the second formula contains "$\Lambda(d,\mathfrak{a})$".]

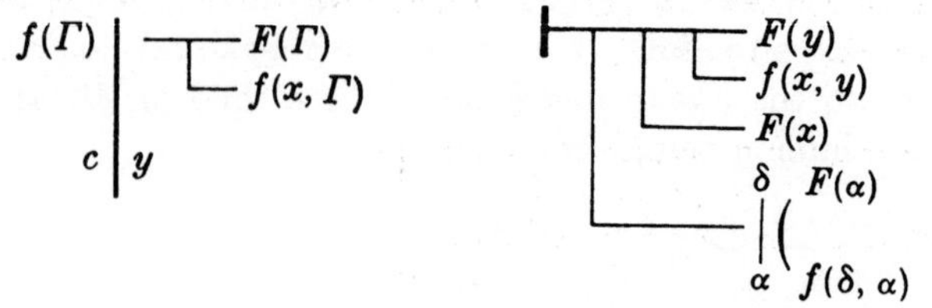

$$\frac{f(\Gamma) \;\Big|\; \begin{array}{l} F(\Gamma) \\ f(x,\Gamma) \end{array}}{c \;\Big|\; y} \qquad \vdash \begin{array}{l} F(y) \\ f(x,y) \\ F(x) \\ \underset{\alpha}{\overset{\delta}{\rightsquigarrow}} \bigl(\begin{array}{l} F(\alpha) \\ f(\delta,\alpha) \end{array} \end{array} \tag{72}$$

If property F is hereditary in the f-sequence, if x has property F, and if y is a result of an application of the procedure f to x, then y has property F.

$$72 \qquad \vdash \begin{array}{l} F(y) \\ f(x,y) \\ F(x) \\ \underset{\alpha}{\overset{\delta}{\rightsquigarrow}} \bigl(\begin{array}{l} F(\alpha) \\ f(\delta,\alpha) \end{array} \end{array}$$

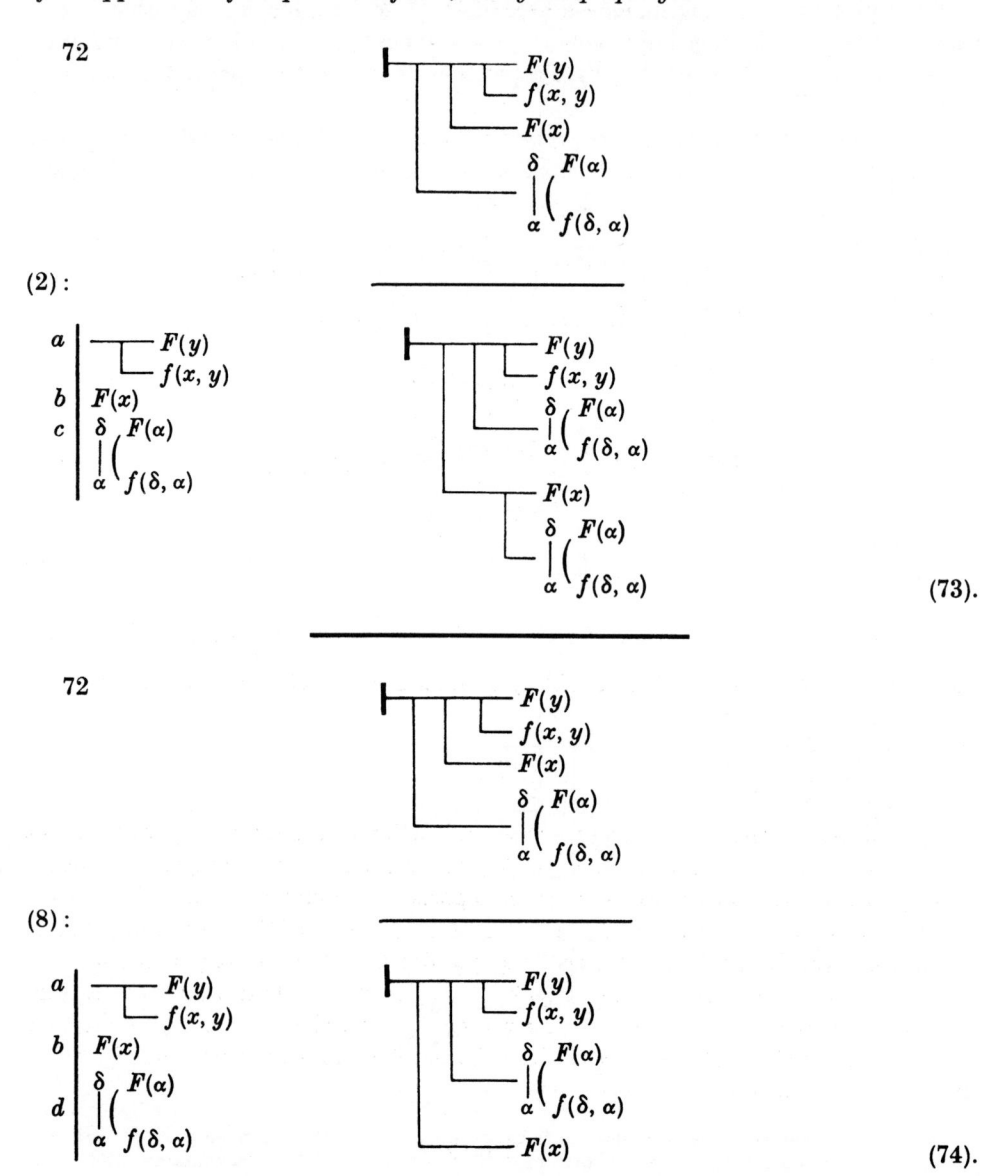

$$(2): \qquad \begin{array}{c|l} a & \begin{array}{l} F(y) \\ f(x,y) \end{array} \\ b & F(x) \\ c & \underset{\alpha}{\overset{\delta}{\rightsquigarrow}} \bigl(\begin{array}{l} F(\alpha) \\ f(\delta,\alpha) \end{array} \end{array} \qquad \vdash \begin{array}{l} F(y) \\ f(x,y) \\ \underset{\alpha}{\overset{\delta}{\rightsquigarrow}} \bigl(\begin{array}{l} F(\alpha) \\ f(\delta,\alpha) \end{array} \\ F(x) \\ \underset{\alpha}{\overset{\delta}{\rightsquigarrow}} \bigl(\begin{array}{l} F(\alpha) \\ f(\delta,\alpha) \end{array} \end{array} \tag{73}$$

$$72 \qquad \vdash \begin{array}{l} F(y) \\ f(x,y) \\ F(x) \\ \underset{\alpha}{\overset{\delta}{\rightsquigarrow}} \bigl(\begin{array}{l} F(\alpha) \\ f(\delta,\alpha) \end{array} \end{array}$$

$$(8): \qquad \begin{array}{c|l} a & \begin{array}{l} F(y) \\ f(x,y) \end{array} \\ b & F(x) \\ d & \underset{\alpha}{\overset{\delta}{\rightsquigarrow}} \bigl(\begin{array}{l} F(\alpha) \\ f(\delta,\alpha) \end{array} \end{array} \qquad \vdash \begin{array}{l} F(y) \\ f(x,y) \\ \underset{\alpha}{\overset{\delta}{\rightsquigarrow}} \bigl(\begin{array}{l} F(\alpha) \\ f(\delta,\alpha) \end{array} \\ F(x) \end{array} \tag{74}$$

If x has a property F that is hereditary in the f-sequence, then every result of an application of the procedure f to x has property F.

$$69 \quad \vdash \left[\left[\overset{\mathfrak{b}}{\smile}\overset{\mathfrak{a}}{\smile} \begin{array}{l} F(\mathfrak{a}) \\ f(\mathfrak{b}, \mathfrak{a}) \\ F(\mathfrak{b}) \end{array} \right] \equiv \underset{\alpha}{\overset{\delta}{\mid}} \Big(\begin{array}{l} F(\alpha) \\ f(\delta, \alpha) \end{array} \right]$$

(52):

$$\begin{array}{c|c} c & \overset{\mathfrak{b}}{\smile}\overset{\mathfrak{a}}{\smile}\begin{array}{l}F(\mathfrak{a})\\f(\mathfrak{b},\mathfrak{a})\\F(\mathfrak{b})\end{array} \\ d & \underset{\alpha}{\overset{\delta}{\mid}}\Big(\begin{array}{l}F(\alpha)\\f(\delta,\alpha)\end{array} \\ f(\Gamma) & \Gamma \end{array} \quad \vdash \begin{array}{l} \underset{\alpha}{\overset{\delta}{\mid}}\Big(\begin{array}{l}F(\alpha)\\f(\delta,\alpha)\end{array}\\ \overset{\mathfrak{b}}{\smile}\overset{\mathfrak{a}}{\smile}\begin{array}{l}F(\mathfrak{a})\\f(\mathfrak{b},\mathfrak{a})\\F(\mathfrak{b})\end{array} \end{array} \quad (75).$$

If from the proposition that $\mathfrak{b}$ has property F, whatever $\mathfrak{b}$ may be, it can be inferred that every result of an application of the procedure f to $\mathfrak{b}$ has property F, then property F is hereditary in the f-sequence.

§ 26.

$$\Vdash \left[\left[\overset{\mathfrak{x}}{\smile}\overset{\mathfrak{a}}{\smile}\begin{array}{l}\mathfrak{F}(y)\\\mathfrak{F}(\mathfrak{a})\\f(x,\mathfrak{a})\end{array} \\ \underset{\alpha}{\overset{\delta}{\mid}}\Big(\begin{array}{l}\mathfrak{F}(\alpha)\\f(\delta,\alpha)\end{array} \right] \equiv \underset{\beta}{\overset{\gamma}{\sim}}f(x_\gamma, y_\beta) \right]$$

(76).

This is the definition of the combination of signs on the right, $\underset{\beta}{\overset{\gamma}{\sim}} f(x_\gamma, y_\beta)$. I refer the reader to § 24 for the use of the double judgment stroke and Greek letters. It would not do to write merely

$$\underset{y}{\overset{x}{\sim}} f(x, y)$$

instead of the expression above since, when a function of x and y is fully written out, these letters could still appear outside of the argument places; in that case we should not be able to tell which places were to be regarded as argument places. Hence these must be characterized as such. This is done here by means of the subscripts γ and β. These must be chosen different since it is possible that the two arguments may be identical with each other. We use Greek letters for this, so that we have a certain freedom of choice and thus can choose the symbols for the argument places of the enclosed expression different from those [used for the argument places] of the enclosing expression in case

$$\underset{\beta}{\overset{\gamma}{\sim}} f(x_\gamma, y_\beta)$$

should enclose within itself a similarly constructed expression. *Whether Greek letters are identical or different is of significance here only within the expression*

$$\overset{\gamma}{\underset{\beta}{\smile}} f(x_\gamma, y_\beta);$$

outside, the same letters could be used, and this would not indicate any connection with the occurrences inside.

We translate

$$\overset{\gamma}{\underset{\beta}{\smile}} f(x_\gamma, y_\beta)$$

by "*y follows x in the f-sequence*", a way of speaking that, to be sure, is possible only when the function *f* is determined. Accordingly, (76) can be rendered in words somewhat as follows:

If from the two propositions that every result of an application of the procedure f to x has property F and that property F is hereditary in the f-sequence, it can be inferred, whatever F may be, that y has property F, then I say: "y follows x in the f-sequence", or "x precedes y in the f-sequence".[19]

§ 27.

76

(68):

(77).

Here $F(y)$, $F(\mathfrak{a})$, and $F(\alpha)$ must be regarded, in accordance with § 10, as different functions of the argument *F*. (77) means:

If y follows x in the f-sequence, if property F is hereditary in the f-sequence, and if every result of an application of the procedure f to x has property F, then y has property F.

---

[19] To make clearer the generality of the concept, given hereby, of succession in a sequence, I remind the reader of a number of possibilities. Not only juxtaposition, such as pearls on a string exhibit, is subsumed here, but also branching like that of a family tree, merging of several branches, and ringlike self-linking.

BEGRIFFSSCHRIFT

77

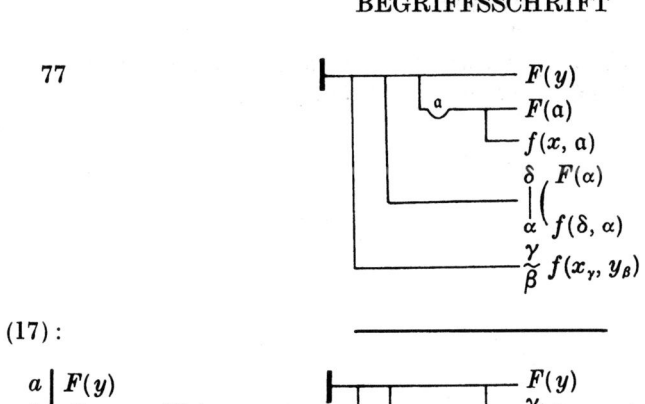

$$\vdash \begin{array}{l} F(y) \\ \mathop{\smile}\limits^{\mathfrak{a}} F(\mathfrak{a}) \\ \phantom{\smile^{\mathfrak{a}}} f(x, \mathfrak{a}) \\ \mathop{|}\limits_{\alpha}^{\delta} \Big(\begin{array}{l} F(\alpha) \\ f(\delta, \alpha) \end{array} \\ \mathop{\gamma}\limits_{\widetilde{\beta}} f(x_\gamma, y_\beta) \end{array}$$

(17):

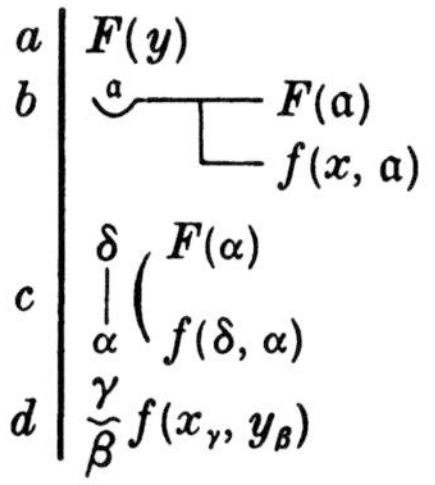

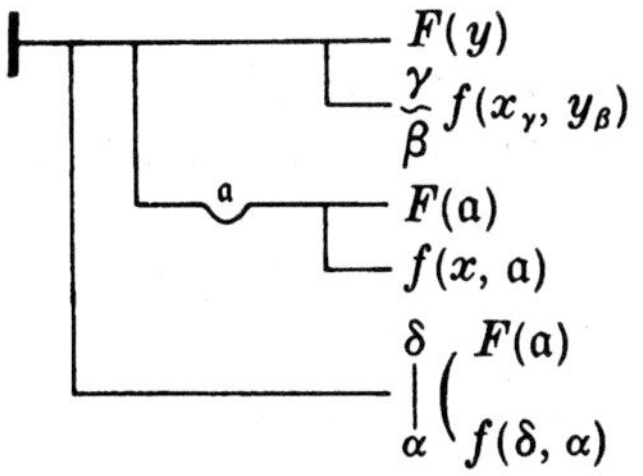

$$\begin{array}{c|l} a & F(y) \\ b & \mathop{\smile}\limits^{\mathfrak{a}} F(\mathfrak{a}) \\ & \phantom{\smile^{\mathfrak{a}}} f(x, \mathfrak{a}) \\ c & \mathop{|}\limits_{\alpha}^{\delta} \Big(\begin{array}{l} F(\alpha) \\ f(\delta, \alpha) \end{array} \\ d & \mathop{\gamma}\limits_{\widetilde{\beta}} f(x_\gamma, y_\beta) \end{array}$$

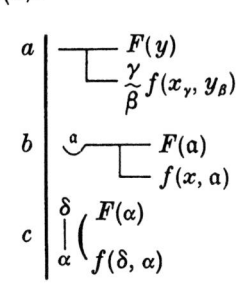

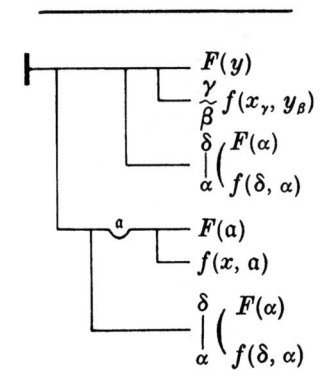

$$\vdash \begin{array}{l} F(y) \\ \mathop{\gamma}\limits_{\widetilde{\beta}} f(x_\gamma, y_\beta) \\ \mathop{\smile}\limits^{\mathfrak{a}} F(\mathfrak{a}) \\ \phantom{\smile^{\mathfrak{a}}} f(x, \mathfrak{a}) \\ \mathop{|}\limits_{\alpha}^{\delta} \Big(\begin{array}{l} F(\alpha) \\ f(\delta, \alpha) \end{array} \end{array}$$ (78).

(2):

$$\begin{array}{c|l} a & F(y) \\ & \mathop{\gamma}\limits_{\widetilde{\beta}} f(x_\gamma, y_\beta) \\ b & \mathop{\smile}\limits^{\mathfrak{a}} F(\mathfrak{a}) \\ & \phantom{\smile^{\mathfrak{a}}} f(x, \mathfrak{a}) \\ c & \mathop{|}\limits_{\alpha}^{\delta} \Big(\begin{array}{l} F(\alpha) \\ f(\delta, \alpha) \end{array} \end{array}$$

$$\vdash \begin{array}{l} F(y) \\ \mathop{\gamma}\limits_{\widetilde{\beta}} f(x_\gamma, y_\beta) \\ \mathop{|}\limits_{\alpha}^{\delta} \Big(\begin{array}{l} F(\alpha) \\ f(\delta, \alpha) \end{array} \\ \mathop{\smile}\limits^{\mathfrak{a}} F(\mathfrak{a}) \\ \phantom{\smile^{\mathfrak{a}}} f(x, \mathfrak{a}) \\ \mathop{|}\limits_{\alpha}^{\delta} \Big(\begin{array}{l} F(\alpha) \\ f(\delta, \alpha) \end{array} \end{array}$$ (79).

(5):

$$\begin{array}{c|l} a & F(y) \\ & \mathop{\gamma}\limits_{\widetilde{\beta}} f(x_\gamma, y_\beta) \\ & \mathop{|}\limits_{\alpha}^{\delta} \Big(\begin{array}{l} F(\alpha) \\ f(\delta, \alpha) \end{array} \\ b & \mathop{\smile}\limits^{\mathfrak{a}} F(\mathfrak{a}) \\ & \phantom{\smile^{\mathfrak{a}}} f(x, \mathfrak{a}) \\ & \mathop{|}\limits_{\alpha}^{\delta} \Big(\begin{array}{l} F(\alpha) \\ f(\delta, \alpha) \end{array} \\ c & F(x) \end{array}$$

$$\vdash \begin{array}{l} F(y) \\ \mathop{\gamma}\limits_{\widetilde{\beta}} f(x_\gamma, y_\beta) \\ \mathop{|}\limits_{\alpha}^{\delta} \Big(\begin{array}{l} F(\alpha) \\ f(\delta, \alpha) \end{array} \\ F(x) \\ \mathop{\smile}\limits^{\mathfrak{a}} F(\mathfrak{a}) \\ \phantom{\smile^{\mathfrak{a}}} f(x, \mathfrak{a}) \\ \mathop{|}\limits_{\alpha}^{\delta} \Big(\begin{array}{l} F(\alpha) \\ f(\delta, \alpha) \end{array} \\ F(x) \end{array}$$ (80).

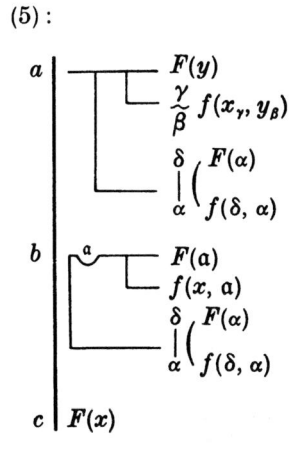

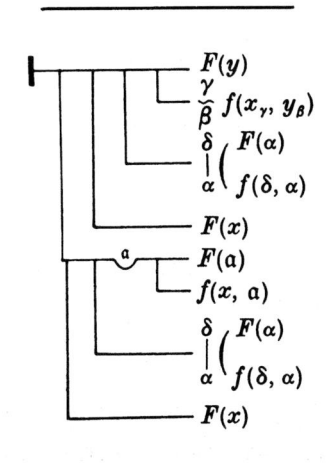

(74) ::

$y \mid \mathfrak{a}$

$$\begin{array}{l} \vdash \!\!\begin{array}{l} \!\!\begin{array}{l} F(y) \\ \dfrac{\gamma}{\beta} f(x_\gamma, y_\beta) \end{array} \\ \!\!\begin{array}{l} \delta \\ \mid \\ \alpha \end{array}\!\!\bigg(\!\!\begin{array}{l} F(\alpha) \\ f(\delta, \alpha) \end{array} \\ F(x) \end{array} \end{array}$$

(81).

Since in (74) y occurs only in

$$\vdash \begin{array}{l} F(y) \\ f(x, y), \end{array}$$

the concavity can, according to § 11, immediately precede this expression, provided y is replaced by the German letter $\mathfrak{a}$. We can translate (81) thus:

If x has a property F that is hereditary in the f-sequence, and if y follows x in the f-sequence, then y has property F.[20]

For example, let F be the property of being a heap of beans; let f be the procedure of removing one bean from a heap of beans; so that $f(a, b)$ means the circumstance that b contains all beans of the heap a except one and does not contain anything else. Then by means of our proposition we would arrive at the result that a single bean, or even none at all, is a heap of beans if the property of being a heap of beans is hereditary in the f-sequence. This is not the case in general, however, since there are certain z for which $F(z)$ cannot become a judgment on account of the indeterminateness of the notion "heap".

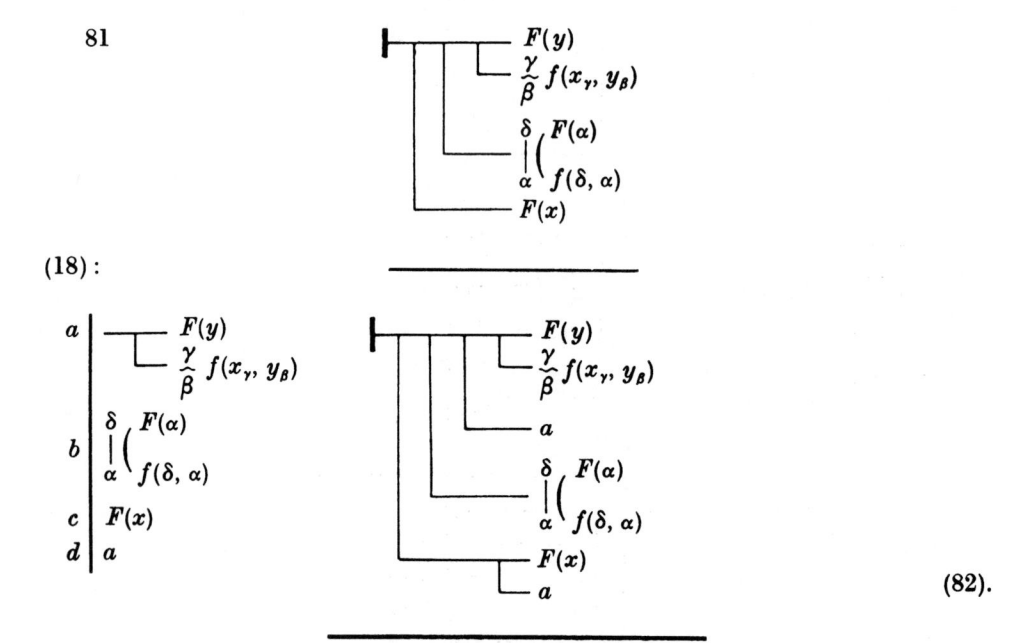

81

$$\vdash \!\!\begin{array}{l} F(y) \\ \dfrac{\gamma}{\beta} f(x_\gamma, y_\beta) \\ \begin{array}{l} \delta \\ \mid \\ \alpha \end{array}\!\!\bigg(\!\!\begin{array}{l} F(\alpha) \\ f(\delta, \alpha) \end{array} \\ F(x) \end{array}$$

(18):

$$\begin{array}{l} a \\ b \\ c \\ d \end{array} \bigg| \begin{array}{l} F(y) \\ \dfrac{\gamma}{\beta} f(x_\gamma, y_\beta) \\ \begin{array}{l} \delta \\ \mid \\ \alpha \end{array}\!\!\bigg(\!\!\begin{array}{l} F(\alpha) \\ f(\delta, \alpha) \end{array} \\ F(x) \\ \mathfrak{a} \end{array} \qquad \vdash \!\!\begin{array}{l} F(y) \\ \dfrac{\gamma}{\beta} f(x_\gamma, y_\beta) \\ \mathfrak{a} \\ \begin{array}{l} \delta \\ \mid \\ \alpha \end{array}\!\!\bigg(\!\!\begin{array}{l} F(\alpha) \\ f(\delta, \alpha) \end{array} \\ F(x) \\ \mathfrak{a} \end{array}$$

(82).

---

[20] Bernoulli's induction rests upon this. [[Jakob Bernoulli is considered one of the originators of mathematical induction, which he used from 1686 on (see *Bernoulli 1686*).]]

BEGRIFFSSCHRIFT 63

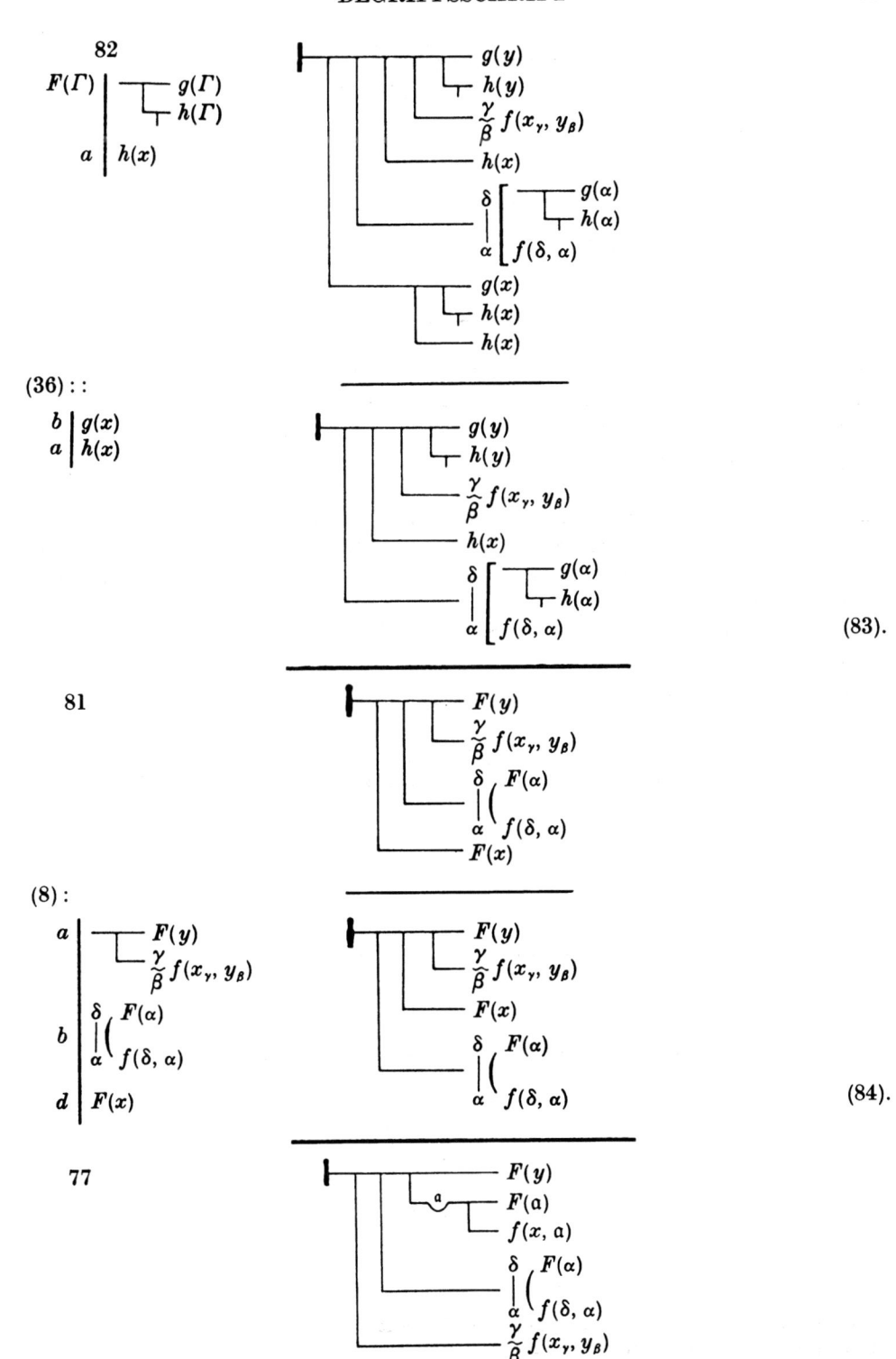

82

$F(\Gamma)$ | ⌐ $g(\Gamma)$
       ⌐ $h(\Gamma)$
a | $h(x)$

├─┬─ $g(y)$
  │ └ $h(y)$
  ├─ $\overset{\gamma}{\underset{\beta}{\sim}} f(x_\gamma, y_\beta)$
  ├─ $h(x)$
  ├─ $\overset{\delta}{\underset{\alpha}{|}} \Big[\begin{matrix} \text{┬ } g(\alpha) \\ \text{└ } h(\alpha) \\ f(\delta, \alpha) \end{matrix}$
  ├─ $g(x)$
  ├─ $h(x)$
  └─ $h(x)$

(36) : :

b | $g(x)$
a | $h(x)$

├─┬─ $g(y)$
  │ └ $h(y)$
  ├─ $\overset{\gamma}{\underset{\beta}{\sim}} f(x_\gamma, y_\beta)$
  ├─ $h(x)$
  └─ $\overset{\delta}{\underset{\alpha}{|}} \Big[\begin{matrix} \text{┬ } g(\alpha) \\ \text{└ } h(\alpha) \\ f(\delta, \alpha) \end{matrix}$

(83).

81

├─ $F(y)$
├─ $\overset{\gamma}{\underset{\beta}{\sim}} f(x_\gamma, y_\beta)$
├─ $\overset{\delta}{\underset{\alpha}{|}} \Big(\begin{matrix} F(\alpha) \\ f(\delta, \alpha) \end{matrix}$
└─ $F(x)$

(8) :

a | ⌐ $F(y)$
      └ $\overset{\gamma}{\underset{\beta}{\sim}} f(x_\gamma, y_\beta)$
b | $\overset{\delta}{\underset{\alpha}{|}} \Big(\begin{matrix} F(\alpha) \\ f(\delta, \alpha) \end{matrix}$
d | $F(x)$

├─ $F(y)$
├─ $\overset{\gamma}{\underset{\beta}{\sim}} f(x_\gamma, y_\beta)$
├─ $F(x)$
└─ $\overset{\delta}{\underset{\alpha}{|}} \Big(\begin{matrix} F(\alpha) \\ f(\delta, \alpha) \end{matrix}$

(84).

77

├─ $F(y)$
├─∪ₐ─ $F(\mathfrak{a})$
│ $f(x, \mathfrak{a})$
├─ $\overset{\delta}{\underset{\alpha}{|}} \Big(\begin{matrix} F(\alpha) \\ f(\delta, \alpha) \end{matrix}$
└─ $\overset{\gamma}{\underset{\beta}{\sim}} f(x_\gamma, y_\beta)$

(12) :

64   FREGE

$$a \;\bigg|\; F(y)$$

$$b \;\bigg|\; \smile^{\mathfrak{a}} \begin{array}{l} F(\mathfrak{a}) \\ f(x, \mathfrak{a}) \end{array}$$

$$c \;\bigg|\; \underset{\alpha}{\overset{\delta}{|}} \bigg(\begin{array}{l} F(\alpha) \\ f(\delta, \alpha) \end{array}$$

$$d \;\bigg|\; \underset{\beta}{\overset{\gamma}{\sim}} f(x_\gamma, y_\beta)$$

$$\vdash \begin{array}{l} F(y) \\ \underset{\alpha}{\overset{\delta}{|}} \bigg(\begin{array}{l} F(\alpha) \\ f(\delta, \alpha) \end{array} \\ \smile^{\mathfrak{a}} \begin{array}{l} F(\mathfrak{a}) \\ f(x, \mathfrak{a}) \end{array} \\ \underset{\beta}{\overset{\gamma}{\sim}} f(x_\gamma, y_\beta) \end{array} \qquad (85).$$

(19):

$$b \;\bigg|\; \begin{array}{l} F(y) \\ \underset{\alpha}{\overset{\delta}{|}} \bigg(\begin{array}{l} F(\alpha) \\ f(\delta, \alpha) \end{array} \end{array}$$

$$c \;\bigg|\; \smile^{\mathfrak{a}} \begin{array}{l} F(\mathfrak{a}) \\ f(x, \mathfrak{a}) \end{array}$$

$$d \;\bigg|\; \underset{\beta}{\overset{\gamma}{\sim}} f(x_\gamma, y_\beta)$$

$$a \;\bigg|\; \begin{array}{l} F(z) \\ f(y, z) \\ \underset{\alpha}{\overset{\delta}{|}} \bigg(\begin{array}{l} F(\alpha) \\ f(\delta, \alpha) \end{array} \end{array}$$

$$\vdash \begin{array}{l} F(z) \\ f(y, z) \\ \underset{\alpha}{\overset{\delta}{|}} \bigg(\begin{array}{l} F(\alpha) \\ f(\delta, \alpha) \end{array} \\ \smile^{\mathfrak{a}} \begin{array}{l} F(\mathfrak{a}) \\ f(x, \mathfrak{a}) \end{array} \\ \underset{\beta}{\overset{\gamma}{\sim}} f(x_\gamma, y_\beta) \\ F(y) \\ \underset{\alpha}{\overset{\delta}{|}} \bigg(\begin{array}{l} F(\alpha) \\ f(\delta, \alpha) \end{array} \end{array} \qquad (86).$$

(73)::

$$\begin{array}{c|c} y & z \\ x & y \end{array}$$

$$\vdash \begin{array}{l} F(z) \\ f(y, z) \\ \underset{\alpha}{\overset{\delta}{|}} \bigg(\begin{array}{l} F(\alpha) \\ f(\delta, \alpha) \end{array} \\ \smile^{\mathfrak{a}} \begin{array}{l} F(\mathfrak{a}) \\ f(x, \mathfrak{a}) \end{array} \\ \underset{\beta}{\overset{\gamma}{\sim}} f(x_\gamma, y_\beta) \end{array} \qquad (87).$$

In words, the derivation of this proposition will be somewhat as follows. Assume that
   (α) y follows x in the f-sequence,
   (β) Every result of an application of the procedure f to x has property F, and
   (γ) Property F is hereditary in the f-sequence.

From these assumptions it follows according to (85) that

(δ) y has property F.

Now,

(ε) Let z be a result of an application of the procedure f to y.

Then by (72) it follows from (γ), (δ), and (ε) that z has property F. Therefore,

If z is a result of an application of the procedure f to an object y that follows x in the f-sequence and if every result of an application of the procedure f to x has a property F that is hereditary in the f-sequence, then z has this property F.[21]

87

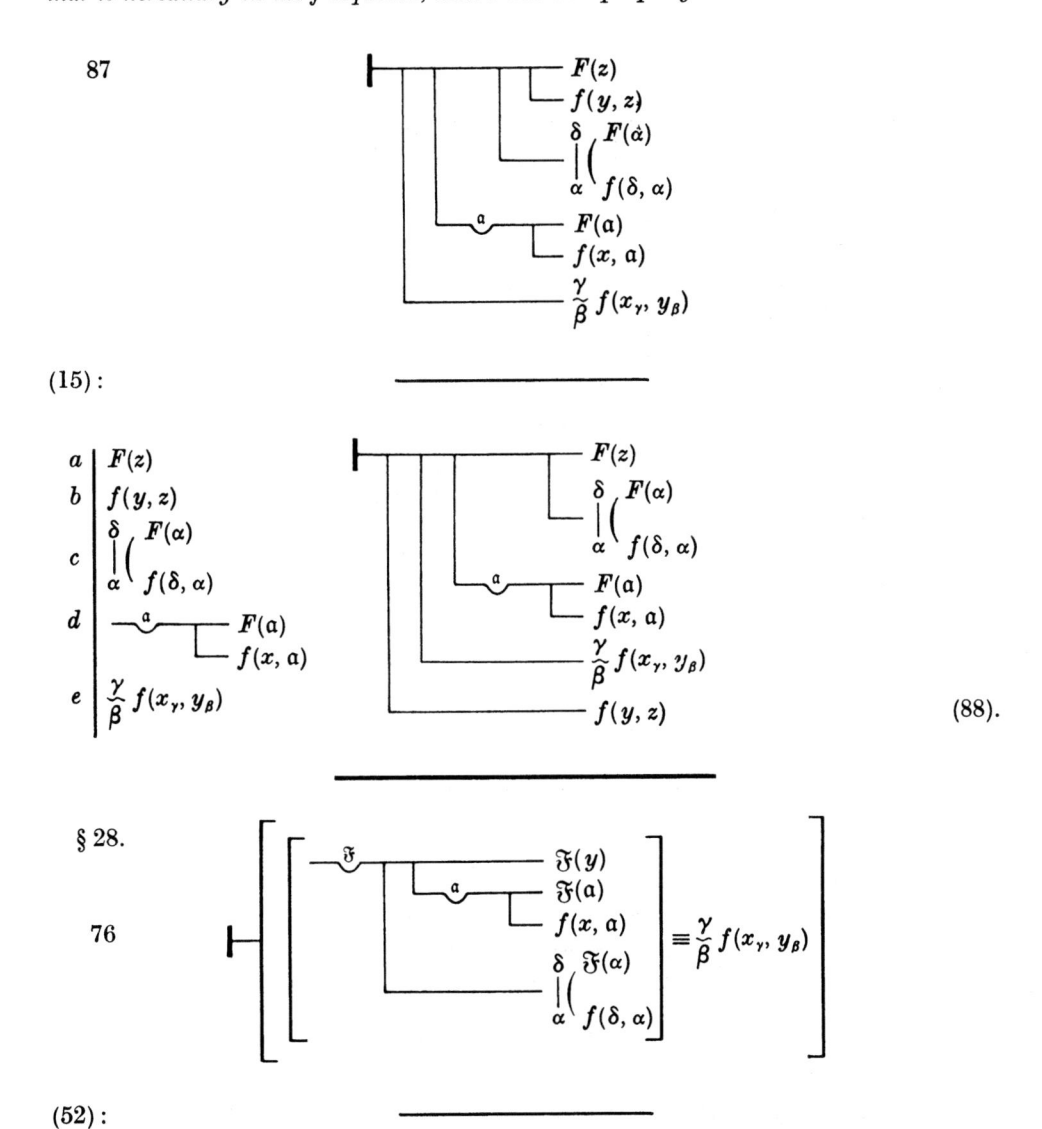

$$\vdash \begin{array}{l} F(z) \\ f(y,z) \\ \underset{\alpha}{\overset{\delta}{\mid}}\!\!\left(\begin{array}{l} F(\dot{\alpha}) \\ f(\delta,\alpha) \end{array}\right. \\ \overset{a}{\frown}\; F(\mathfrak{a}) \\ \qquad f(x,\mathfrak{a}) \\ \underset{\beta}{\overset{\gamma}{\widetilde{}}}\, f(x_\gamma, y_\beta) \end{array}$$

(15):

$$\begin{array}{l|l} a & F(z) \\ b & f(y,z) \\ c & \underset{\alpha}{\overset{\delta}{\mid}}\!\left(\begin{array}{l} F(\alpha) \\ f(\delta,\alpha) \end{array}\right. \\ d & \overset{a}{\frown}\; F(\mathfrak{a}) \\ & \qquad f(x,\mathfrak{a}) \\ e & \underset{\beta}{\overset{\gamma}{\widetilde{}}}\, f(x_\gamma, y_\beta) \end{array} \qquad \vdash \begin{array}{l} F(z) \\ \underset{\alpha}{\overset{\delta}{\mid}}\!\left(\begin{array}{l} F(\alpha) \\ f(\delta,\alpha) \end{array}\right. \\ \overset{a}{\frown}\; F(\mathfrak{a}) \\ \qquad f(x,\mathfrak{a}) \\ \underset{\beta}{\overset{\gamma}{\widetilde{}}}\, f(x_\gamma, y_\beta) \\ f(y,z) \end{array}$$ (88).

§ 28.

76

$$\vdash \left[\left[\underset{}{\overset{\mathfrak{F}}{\frown}}\begin{array}{l} \mathfrak{F}(y) \\ \mathfrak{F}(\mathfrak{a}) \\ f(x,\mathfrak{a}) \\ \underset{\alpha}{\overset{\delta}{\mid}}\!\left(\begin{array}{l} \mathfrak{F}(\alpha) \\ f(\delta,\alpha) \end{array}\right. \end{array}\right] \equiv \underset{\beta}{\overset{\gamma}{\widetilde{}}}\, f(x_\gamma, y_\beta)\right]$$

(52):

---

[21] ⟦At the place that corresponds to the last occurrence of "f" in this sentence the German text mistakenly has "F".⟧

$$
(5): \quad
\begin{array}{c|l}
f(\Gamma) & \Gamma \\
c & \mathfrak{F} \begin{array}{l} \mathfrak{F}(y) \\ {}_a\!\!\!\!\!\!\lfloor \mathfrak{F}(\mathfrak{a}) \\ f(x,\mathfrak{a}) \end{array} \\
 & \phantom{\mathfrak{F}}{}^{\delta}\!\!\diagup\!\!\mathfrak{F}(\alpha) \\
 & \phantom{\mathfrak{F}}{}_{\alpha}\!\!\diagdown\!\!f(\delta,\alpha) \\
d & \dfrac{\gamma}{\beta} f(x_\gamma, y_\beta)
\end{array}
\quad\vdash
\begin{array}{l}
\dfrac{\gamma}{\beta} f(x_\gamma, y_\beta) \\
\mathfrak{F} \begin{array}{l} \mathfrak{F}(y) \\ {}_a\!\!\lfloor \mathfrak{F}(\mathfrak{a}) \\ f(x,\mathfrak{a}) \end{array} \\
{}^{\delta}\!\!\diagup\!\!\mathfrak{F}(\alpha) \\
{}_{\alpha}\!\!\diagdown\!\!f(\delta,\alpha)
\end{array}
\quad (89).
$$

$$
\begin{array}{c|l}
a & \dfrac{\gamma}{\beta} f(x_\gamma, y_\beta) \\
b & \mathfrak{F} \begin{array}{l} \mathfrak{F}(y) \\ {}_a\!\!\lfloor \mathfrak{F}(\mathfrak{a}) \\ f(x,\mathfrak{a}) \end{array} \\
 & {}^{\delta}\!\!\diagup\!\!\mathfrak{F}(\alpha) \\
 & {}_{\alpha}\!\!\diagdown\!\!f(\delta,\alpha)
\end{array}
\quad\vdash
\begin{array}{l}
\dfrac{\gamma}{\beta} f(x_\gamma, y_\beta) \\
c \\
\mathfrak{F} \begin{array}{l} \mathfrak{F}(y) \\ {}_a\!\!\lfloor \mathfrak{F}(\mathfrak{a}) \\ f(x,\mathfrak{a}) \end{array} \\
{}^{\delta}\!\!\diagup\!\!\mathfrak{F}(\alpha) \\
{}_{\alpha}\!\!\diagdown\!\!f(\delta,\alpha) \\
c
\end{array}
\quad (90).
$$

$$
63 \quad
\begin{array}{c|l}
f & \mathfrak{F} \\
x & y \\
g(\Gamma) & f(x,\Gamma) \\
m & {}^{\delta}\!\!\diagup\!\!\mathfrak{F}(\alpha) \\
 & {}_{\alpha}\!\!\diagdown\!\!f(\delta,\alpha)
\end{array}
\quad\vdash
\begin{array}{l}
\mathfrak{F} \begin{array}{l} \mathfrak{F}(y)^{22} \\ {}_a\!\!\lfloor \mathfrak{F}(\mathfrak{a}) \\ f(x,\mathfrak{a}) \end{array} \\
{}^{\delta}\!\!\diagup\!\!\mathfrak{F}(\alpha) \\
{}_{\alpha}\!\!\diagdown\!\!f(\delta,\alpha) \\
f(x,y)
\end{array}
$$

$$
(90): \quad
\begin{array}{c|l}
c & f(x,y)
\end{array}
\quad\vdash
\begin{array}{l}
\dfrac{\gamma}{\beta} f(x_\gamma, y_\beta) \\
f(x,y)
\end{array}
\quad (91).
$$

Let us give here the derivation of proposition (91) in words. From the proposition (α), "Every result of an application of the procedure f to x has property $\mathfrak{F}$", it can be inferred, whatever $\mathfrak{F}$ may be, that every result of an application of the procedure f to x has property $\mathfrak{F}$. Hence it can also be inferred from proposition (α) and the proposition that property $\mathfrak{F}$ is hereditary in the f-sequence, whatever $\mathfrak{F}$ may be, that every result of an application of the procedure f to x has property $\mathfrak{F}$.

Therefore, according to (90) the following proposition holds:

Every result of an application of a procedure f to an object x follows that x in the f-sequence.

---

[22] Concerning the concavity with $\mathfrak{F}$ see § 11. ⟦In fact, Frege has already used the concavity with $\mathfrak{F}$ several times, the first occurrence being in (76).⟧

BEGRIFFSSCHRIFT 67

91
$$\vdash\!\!\begin{array}{l}-\underset{\beta}{\overset{\gamma}{\sim}}f(x_\gamma, y_\beta)\\ -f(x, y)\end{array}$$

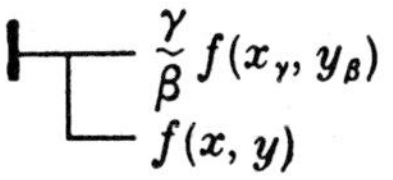

(53):

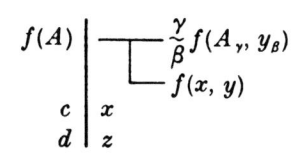

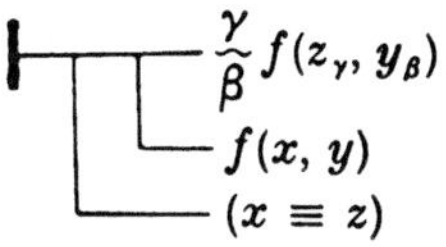

$$\begin{array}{c|c} f(A) & \vdash\!\!\begin{array}{l}-\underset{\beta}{\overset{\gamma}{\sim}}f(A_\gamma, y_\beta)\\ -f(x, y)\end{array} \\ c & x \\ d & z \end{array}$$

$$\vdash\!\!\begin{array}{l}-\underset{\beta}{\overset{\gamma}{\sim}}f(z_\gamma, y_\beta)\\ -f(x, y)\\ -(x\equiv z)\end{array}$$

(92).

$$\begin{array}{c|c} 60 & \\ & \mathfrak{a} \mid \mathfrak{F} \\ f(\Gamma) & \Gamma(y) \\ g(\Gamma) & \begin{array}{l}\delta\\ \mid\end{array}\!\!\begin{pmatrix}\Gamma(\alpha)\\ f(\delta,\alpha)\end{pmatrix} \\ h(\Gamma) & \overset{\mathfrak{a}}{\smile}\!\!\begin{array}{l}-\Gamma(\mathfrak{a})\\ -f(x,\mathfrak{a})\end{array} \\ b & \mathfrak{F} \end{array}$$

$$\vdash\!\!\begin{array}{l}\mathfrak{F}\!\!-\!\mathfrak{F}(y)\\ \phantom{\mathfrak{F}}\overset{\mathfrak{a}}{\smile}\mathfrak{F}(\mathfrak{a})\\ \phantom{\mathfrak{F}}-f(x,\mathfrak{a})\\ \phantom{\mathfrak{F}}\begin{array}{l}\delta\\ \mid\end{array}\!\!\begin{pmatrix}\mathfrak{F}(\alpha)\\ f(\delta,\alpha)\end{pmatrix}\\ \mathfrak{F}\!\!-\!\mathfrak{F}(y)\\ \phantom{\mathfrak{F}}\begin{array}{l}\delta\\ \mid\end{array}\!\!\begin{pmatrix}\mathfrak{F}(\alpha)\\ f(\delta,\alpha)\end{pmatrix}\\ \phantom{\mathfrak{F}}\overset{\mathfrak{a}}{\smile}\mathfrak{F}(\mathfrak{a})\\ \phantom{\mathfrak{F}}-f(x,\mathfrak{a})\end{array}$$

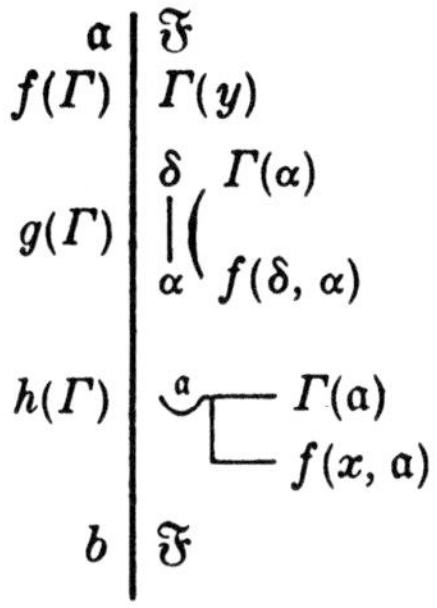

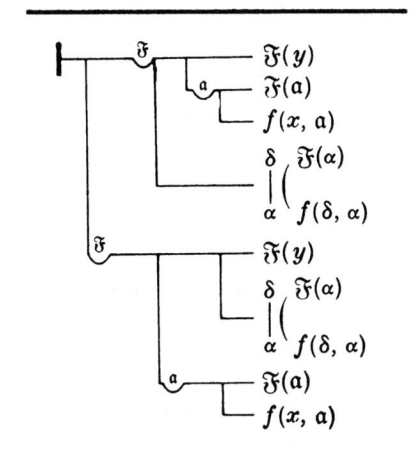

(90):

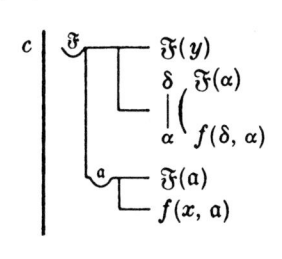

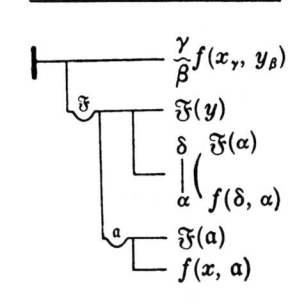

$$\begin{array}{c|l} c & \mathfrak{F}\!\!-\!\mathfrak{F}(y)\\ & \phantom{\mathfrak{F}}\begin{array}{l}\delta\\ \mid\end{array}\!\!\begin{pmatrix}\mathfrak{F}(\alpha)\\ f(\delta,\alpha)\end{pmatrix}\\ & \phantom{\mathfrak{F}}\overset{\mathfrak{a}}{\smile}\mathfrak{F}(\mathfrak{a})\\ & \phantom{\mathfrak{F}}-f(x,\mathfrak{a}) \end{array}$$

$$\vdash\!\!\begin{array}{l}-\underset{\beta}{\overset{\gamma}{\sim}}f(x_\gamma, y_\beta)\\ \mathfrak{F}\!\!-\!\mathfrak{F}(y)\\ \phantom{\mathfrak{F}}\begin{array}{l}\delta\\ \mid\end{array}\!\!\begin{pmatrix}\mathfrak{F}(\alpha)\\ f(\delta,\alpha)\end{pmatrix}\\ \phantom{\mathfrak{F}}\overset{\mathfrak{a}}{\smile}\mathfrak{F}(\mathfrak{a})\\ \phantom{\mathfrak{F}}-f(x,\mathfrak{a})\end{array}$$

(93).

93
$$y \mid z$$

$$\vdash\!\!\begin{array}{l}-\underset{\beta}{\overset{\gamma}{\sim}}f(x_\gamma, z_\beta)\\ \mathfrak{F}\!\!-\!\mathfrak{F}(z)\\ \phantom{\mathfrak{F}}\begin{array}{l}\delta\\ \mid\end{array}\!\!\begin{pmatrix}\mathfrak{F}(\alpha)\\ f(\delta,\alpha)\end{pmatrix}\\ \phantom{\mathfrak{F}}\overset{\mathfrak{a}}{\smile}\mathfrak{F}(\mathfrak{a})\\ \phantom{\mathfrak{F}}-f(x,\mathfrak{a})\end{array}$$

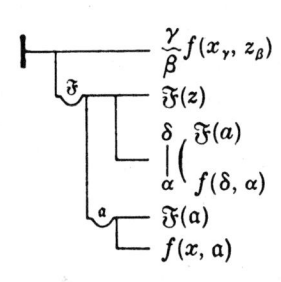

(7):

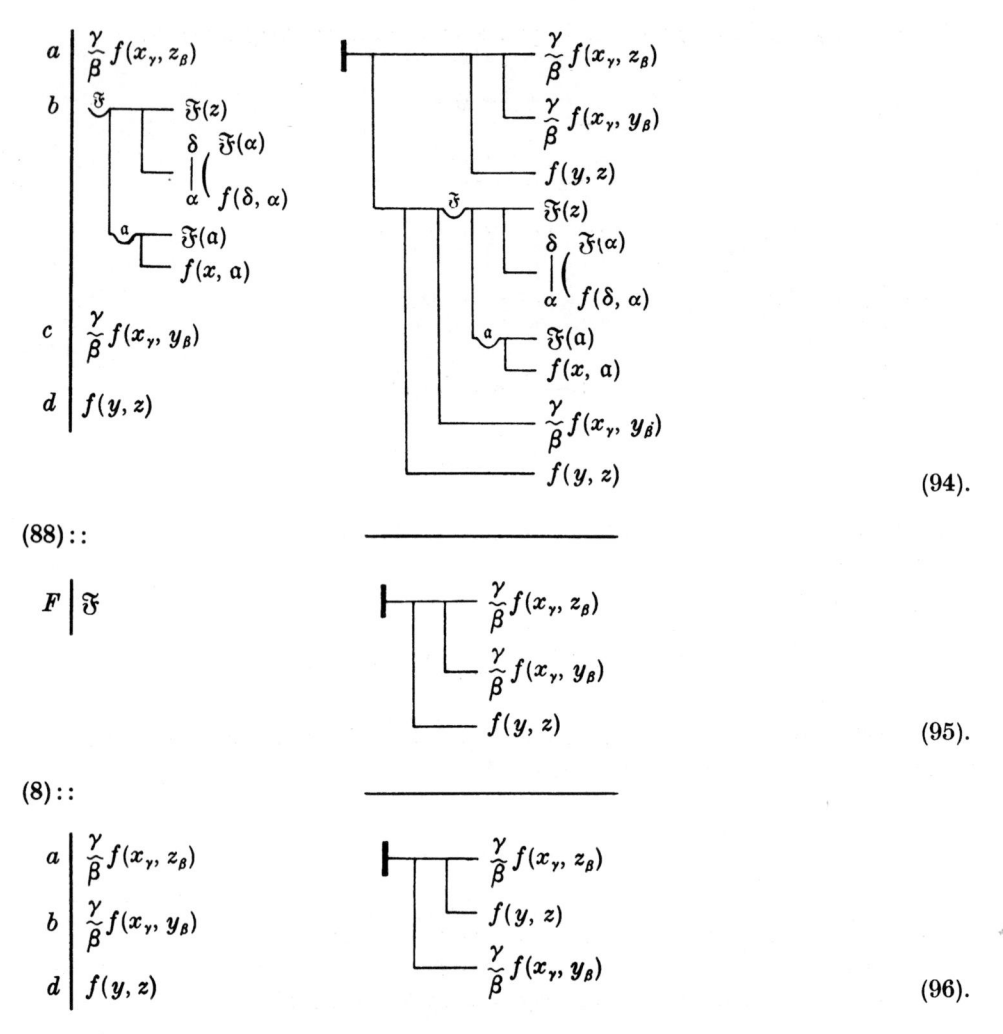

$$a \left| \genfrac{}{}{0pt}{}{\gamma}{\beta} f(x_\gamma, z_\beta) \right.$$

$$b \left| \mathfrak{F} \begin{cases} \mathfrak{F}(z) \\ \genfrac{}{}{0pt}{}{\delta}{\alpha}\bigg(\begin{matrix} \mathfrak{F}(\alpha) \\ f(\delta, \alpha) \end{matrix} \\ \mathfrak{a} \begin{cases} \mathfrak{F}(\mathfrak{a}) \\ f(x, \mathfrak{a}) \end{cases} \end{cases} \right.$$

$$c \left| \genfrac{}{}{0pt}{}{\gamma}{\beta} f(x_\gamma, y_\beta) \right.$$

$$d \left| f(y, z) \right.$$

$$\vdash \begin{cases} \genfrac{}{}{0pt}{}{\gamma}{\beta} f(x_\gamma, z_\beta) \\ \genfrac{}{}{0pt}{}{\gamma}{\beta} f(x_\gamma, y_\beta) \\ f(y, z) \\ \mathfrak{F} \begin{cases} \mathfrak{F}(z) \\ \genfrac{}{}{0pt}{}{\delta}{\alpha}\bigg(\begin{matrix} \mathfrak{F}(\alpha) \\ f(\delta, \alpha) \end{matrix} \\ \mathfrak{a} \begin{cases} \mathfrak{F}(\mathfrak{a}) \\ f(x, \mathfrak{a}) \end{cases} \end{cases} \\ \genfrac{}{}{0pt}{}{\gamma}{\beta} f(x_\gamma, y_\beta) \\ f(y, z) \end{cases} \tag{94}.$$

(88)::

$$F \mid \mathfrak{F}$$

$$\vdash \begin{cases} \genfrac{}{}{0pt}{}{\gamma}{\beta} f(x_\gamma, z_\beta) \\ \genfrac{}{}{0pt}{}{\gamma}{\beta} f(x_\gamma, y_\beta) \\ f(y, z) \end{cases} \tag{95}.$$

(8)::

$$a \left| \genfrac{}{}{0pt}{}{\gamma}{\beta} f(x_\gamma, z_\beta) \right.$$
$$b \left| \genfrac{}{}{0pt}{}{\gamma}{\beta} f(x_\gamma, y_\beta) \right.$$
$$d \left| f(y, z) \right.$$

$$\vdash \begin{cases} \genfrac{}{}{0pt}{}{\gamma}{\beta} f(x_\gamma, z_\beta) \\ f(y, z) \\ \genfrac{}{}{0pt}{}{\gamma}{\beta} f(x_\gamma, y_\beta) \end{cases} \tag{96}.$$

Every result of an application of the procedure f to an object that follows x in the f-sequence follows x in the f-sequence.

96
$z \mid \mathfrak{a}$
$y \mid \mathfrak{b}$

$$\vdash \begin{cases} {}^{\mathfrak{b}}\!\frown\!{}^{\mathfrak{a}}\!\frown\! \genfrac{}{}{0pt}{}{\gamma}{\beta} f(x_\gamma, \mathfrak{a}_\beta) \\ f(\mathfrak{b}, \mathfrak{a}) \\ \genfrac{}{}{0pt}{}{\gamma}{\beta} f(x_\gamma, \mathfrak{b}_\beta) \end{cases}$$

(75):

$$F(\Gamma) \left| \genfrac{}{}{0pt}{}{\gamma}{\beta} f(x_\gamma, \Gamma_\beta) \right.$$

$$\vdash \genfrac{}{}{0pt}{}{\delta}{\alpha}\bigg(\begin{matrix} \genfrac{}{}{0pt}{}{\gamma}{\beta} f(x_\gamma, \alpha_\beta) \\ f(\delta, \alpha) \end{matrix} \tag{97}.$$

The property of following x in the f-sequence is hereditary in the f-sequence.

$$97 \quad \vdash \begin{array}{c} \delta \\ | \\ \alpha \end{array} \Big(\underset{\beta}{\overset{\gamma}{\approx}} f(x_\gamma, \alpha_\beta) \\ f(\delta, \alpha) \Big)$$

(84):

$$\begin{array}{c|c} F(\Gamma) & \underset{\beta}{\overset{\gamma}{\approx}} f(x_\gamma, \Gamma_\beta) \\ x & y \\ y & z \end{array} \qquad \vdash \begin{array}{c} \underset{\beta}{\overset{\gamma}{\approx}} f(x_\gamma, z_\beta) \\ \underset{\beta}{\overset{\gamma}{\approx}} f(y_\gamma, z_\beta) \\ \underset{\beta}{\overset{\gamma}{\approx}} f(x_\gamma, y_\beta) \end{array} \tag{98}$$

If y follows x in the f-sequence and if z follows y in the f-sequence, then z follows x in the f-sequence.

§ 29.

$$\vdash \left[\left[\begin{array}{c} (z \equiv x) \\ \underset{\beta}{\overset{\gamma}{\approx}} f(x_\gamma, z_\beta) \end{array} \right] \equiv \underset{\beta}{\overset{\gamma}{\approx}} f(x_\gamma, z_\beta) \right] \tag{99}$$

Here I refer the reader to what was said about the introduction of new signs in connection with formulas (69) and (76). Let

$$\underset{\beta}{\overset{\gamma}{\approx}} f(x_\gamma, z_\beta)$$

be translated by "*z belongs to the f-sequence beginning with x*" or by "*x belongs to the f-sequence ending with z*". Then in words (99) reads:

If z is identical with x or follows x in the f-sequence, then I say: "z belongs to the f-sequence beginning with x" or "x belongs to the f-sequence ending with z".

$$99 \quad \vdash \left[\left[\begin{array}{c} (z \equiv x) \\ \underset{\beta}{\overset{\gamma}{\approx}} f(x_\gamma, z_\beta) \end{array} \right] \equiv \underset{\beta}{\overset{\gamma}{=}} f(x_\gamma, z_\beta) \right]$$

(57):

$$\begin{array}{c|c} f(\Gamma) & \Gamma \\ c & \begin{array}{c} (z \equiv x) \\ \underset{\beta}{\overset{\gamma}{\approx}} f(x_\gamma, z_\beta) \end{array} \\ d & \underset{\beta}{\overset{\gamma}{=}} f(x_\gamma, z_\beta) \end{array} \qquad \vdash \begin{array}{c} (z \equiv x) \\ \underset{\beta}{\overset{\gamma}{\approx}} f(x_\gamma, z_\beta) \\ \underset{\beta}{\overset{\gamma}{=}} f(x_\gamma, z_\beta) \end{array} \tag{100}$$

(48):

$$
\begin{array}{c|l}
b & (z \equiv x) \\
c & \dfrac{\gamma}{\beta} f(x_\gamma, z_\beta) \\
d & \dfrac{\gamma}{\beta} f(x_\gamma, z_\beta) \\
a & \begin{array}{l} \dfrac{\gamma}{\beta} f(x_\gamma, v_\beta) \\ f(z, v) \end{array}
\end{array}
\qquad
\begin{array}{l}
\vdash \dfrac{\gamma}{\beta} f(x_\gamma, v_\beta) \\
\quad\; f(z, v) \\
\quad\; \dfrac{\gamma}{\beta} f(x_\gamma, z_\beta) \\
\quad\; \dfrac{\gamma}{\beta} f(x_\gamma, v_\beta) \\
\quad\; f(z, v) \\
\quad\; \dfrac{\gamma}{\beta} f(x_\gamma, z_\beta) \\
\quad\; \dfrac{\gamma}{\beta} f(x_\gamma, v_\beta) \\
\quad\; f(z, v) \\
\quad\; (z \equiv x)
\end{array}
\qquad (101).
$$

$$
(96, 92) ::
\begin{array}{c|c|c|c}
y & z & x & z \\
z & v & z & x \\
  &   & y & v
\end{array}
\qquad
\begin{array}{l}
\vdash \dfrac{\gamma}{\beta} f(x_\gamma, v_\beta)^{23} \\
\quad\; f(z, v) \\
\quad\; \dfrac{\gamma}{\beta} f(x_\gamma, z_\beta)
\end{array}
\qquad (102).
$$

Let us here give the derivation of (102) in words.

If z is the same as x, then by (92) every result of an application of the procedure f to z follows x in the f-sequence. If z follows x in the f-sequence, then by (96) every result of an application of f to z follows x in the f-sequence.

From these two propositions it follows, according to (101), that:

If z belongs to the f-sequence beginning with x, then every result of an application of the procedure f to z follows x in the f-sequence.

$$
100 \qquad
\begin{array}{l}
\vdash (z \equiv x) \\
\quad\; \dfrac{\gamma}{\beta} f(x_\gamma, z_\beta) \\
\quad\; \dfrac{\gamma}{\beta} f(x_\gamma, z_\beta)
\end{array}
$$

$$
(19):
\begin{array}{c|l}
b & (z \equiv x) \\
c & \dfrac{\gamma}{\beta} f(x_\gamma, z_\beta) \\
d & \dfrac{\gamma}{\beta} f(x_\gamma, z_\beta) \\
a & (x \equiv z)
\end{array}
\qquad
\begin{array}{l}
\vdash (x \equiv z) \\
\quad\; \dfrac{\gamma}{\beta} f(x_\gamma, z_\beta) \\
\quad\; \dfrac{\gamma}{\beta} f(x_\gamma, z_\beta) \\
\quad\; (x \equiv z) \\
\quad\; (z \equiv x)
\end{array}
\qquad (103).
$$

[23] Concerning the last inference see § 6.

BEGRIFFSSCHRIFT

71

(55) ::

$$\begin{array}{c|c} d & x \\ c & z \end{array}$$

$$\vdash \begin{array}{c} (x \equiv z) \\ \underset{\beta}{\overset{\gamma}{\frown}} f(x_\gamma, z_\beta) \\ \underset{\beta}{\overset{\gamma}{\sim}} f(x_\gamma, z_\beta) \end{array} \qquad (104).$$

§ 30.

99

$$\vdash \left[\left[\begin{array}{c} (z \equiv x) \\ \underset{\beta}{\overset{\gamma}{\frown}} f(x_\gamma, z_\beta) \end{array} \right] \equiv \underset{\beta}{\overset{\gamma}{\sim}} f(x_\gamma, z_\beta) \right]$$

(52) :

$$\begin{array}{c|c} f(\Gamma) & \Gamma \\ c & \begin{array}{c} (z \equiv x) \\ \underset{\beta}{\overset{\gamma}{\frown}} f(x_\gamma, z_\beta) \end{array} \\ d & \underset{\beta}{\overset{\gamma}{\sim}} f(x_\gamma, z_\beta) \end{array}$$

$$\vdash \begin{array}{c} \underset{\beta}{\overset{\gamma}{\sim}} f(x_\gamma, z_\beta) \\ (z \equiv x) \\ \underset{\beta}{\overset{\gamma}{\frown}} f(x_\gamma, z_\beta) \end{array} \qquad (105).$$

(37) :

$$\begin{array}{c|c} a & \underset{\beta}{\overset{\gamma}{\sim}} f(x_\gamma, z_\beta) \\ b & (z \equiv x) \\ c & \underset{\beta}{\overset{\gamma}{\frown}} f(x_\gamma, z_\beta) \end{array}$$

$$\vdash \begin{array}{c} \underset{\beta}{\overset{\gamma}{\frown}} f(x_\gamma, z_\beta) \\ \underset{\beta}{\overset{\gamma}{\sim}} f(x_\gamma, z_\beta) \end{array} \qquad (106).$$

Whatever follows x in the f-sequence belongs to the f-sequence beginning with x.

$$\begin{array}{c} 106 \\ x \mid z \\ z \mid v \end{array}$$

$$\vdash \begin{array}{c} \underset{\beta}{\overset{\gamma}{\sim}} f(z_\gamma, v_\beta) \\ \underset{\beta}{\overset{\gamma}{\frown}} f(z_\gamma, v_\beta) \end{array}$$

(7) :

$$\begin{array}{c|c} a & \underset{\beta}{\overset{\gamma}{\sim}} f(z_\gamma, v_\beta) \\ b & \underset{\beta}{\overset{\gamma}{\frown}} f(z_\gamma, v_\beta) \\ c & f(y, v) \\ d & \underset{\beta}{\overset{\gamma}{\frown}} f(z_\gamma, y_\beta) \end{array}$$

$$\vdash \begin{array}{c} \underset{\beta}{\overset{\gamma}{\sim}} f(z_\gamma, v_\beta) \\ f(y, v) \\ \underset{\beta}{\overset{\gamma}{\frown}} f(z_\gamma, y_\beta) \\ \underset{\beta}{\overset{\gamma}{\sim}} f(z_\gamma, v_\beta) \\ f(y, v) \\ \underset{\beta}{\overset{\gamma}{\sim}} f(z_\gamma, y_\beta) \end{array} \qquad (107).$$

(102) ::

$$\begin{array}{c|c} x & z \\ z & y \end{array}$$

$$\vdash \begin{array}{l} \underset{\beta}{\overset{\gamma}{\approx}} f(z_\gamma, v_\beta) \\ f(y, v) \\ \underset{\beta}{\overset{\gamma}{\approx}} f(z_\gamma, y_\beta) \end{array} \qquad (108).$$

Let us here give the derivation of (108) in words.

If y belongs to the f-sequence beginning with z, then by (102) every result of an application of the procedure f to y follows z in the f-sequence. Then by (106) every result of an application of the procedure f to y belongs to the f-sequence beginning with z. Therefore,

If y belongs to the f-sequence beginning with z, then every result of an application of the procedure f to y belongs to the f-sequence beginning with z.

$$\begin{array}{l} 108 \\ \begin{array}{c|c} v & a \\ z & x \\ y & b \end{array} \end{array} \qquad \vdash \begin{array}{l} \underset{\beta}{\overset{\gamma}{\approx}} f(x_\gamma, a_\beta) \\ f(b, a) \\ \underset{\beta}{\overset{\gamma}{\approx}} f(x_\gamma, b_\beta) \end{array}$$

(75):

$$F(\Gamma) \bigg| \underset{\beta}{\overset{\gamma}{\approx}} f(x_\gamma, \Gamma_\beta) \qquad \vdash \underset{\alpha}{\overset{\delta}{|}} \Big(\underset{\beta}{\overset{\gamma}{\approx}} f(x_\gamma, \alpha_\beta) \atop f(\delta, \alpha) \Big) \qquad (109).$$

The property of belonging to the f-sequence beginning with x is hereditary in the f-sequence.

$$109 \qquad \vdash \underset{\alpha}{\overset{\delta}{|}} \Big(\underset{\beta}{\overset{\gamma}{\approx}} f(x_\gamma, \alpha_\beta) \atop f(\delta, \alpha) \Big)$$

(78):

$$F(\Gamma) \bigg| \underset{\beta}{\overset{\gamma}{\approx}} f(x_\gamma, \Gamma_\beta) \\ \begin{array}{c|c} x & y \\ y & m \end{array} \qquad \vdash \begin{array}{l} \underset{\beta}{\overset{\gamma}{\approx}} f(x_\gamma, m_\beta) \\ \underset{\beta}{\overset{\gamma}{\approx}} f(y_\gamma, m_\beta) \\ \underset{\beta}{\overset{\gamma}{\approx}} f(x_\gamma, a_\beta) \\ f(y, a) \end{array} \qquad (110).$$

$$108 \qquad \vdash \begin{array}{l} \underset{\beta}{\overset{\gamma}{\approx}} f(z_\gamma, v_\beta) \\ f(y, v) \\ \underset{\beta}{\overset{\gamma}{\approx}} f(z_\gamma, y_\beta) \end{array}$$

(25):

BEGRIFFSSCHRIFT

$$
\begin{array}{c|l}
a & \dfrac{\gamma}{\beta} f(z_\gamma, v_\beta) \\
c & f(y, v) \\
d & \dfrac{\gamma}{\beta} f(z_\gamma, y_\beta) \\
b & \vdash \dfrac{\gamma}{\beta} f(v_\gamma, z_\beta)
\end{array}
\qquad
\vdash\!\!\begin{array}{l}
\dfrac{\gamma}{\beta} f(z_\gamma, v_\beta) \\
\dfrac{\gamma}{\beta} f(v_\gamma, z_\beta) \\
f(y, v) \\
\dfrac{\gamma}{\beta} f(z_\gamma, y_\beta)
\end{array}
\qquad (111).
$$

In words the derivation of (111) is as follows:

If y belongs to the f-sequence beginning with z, then by (108) every result of an application of the procedure f to y belongs to the f-sequence beginning with z. Hence every result of an application of the procedure f to y belongs to the f-sequence beginning with z or precedes z in the f-sequence. Therefore,

If y belongs to the f-sequence beginning with z, then every result of an application of the procedure f to y belongs to the f-sequence beginning with z or precedes z in the f-sequence.

105
$$
\vdash\!\!\begin{array}{l}
\dfrac{\gamma}{\beta} f(x_\gamma, z_\beta) \\
(z \equiv x) \\
\dfrac{\gamma}{\beta} f(x_\gamma, z_\beta)
\end{array}
$$

(11):

$$
\begin{array}{c|l}
a & \dfrac{\gamma}{\beta} f(x_\gamma, z_\beta) \\
b & (z \equiv x) \\
c & \vdash \dfrac{\gamma}{\beta} f(x_\gamma, z_\beta)
\end{array}
\qquad
\vdash\!\!\begin{array}{l}
\dfrac{\gamma}{\beta} f(x_\gamma, z_\beta) \\
(z \equiv x)
\end{array}
\qquad (112).
$$

(7):

$$
\begin{array}{c|l}
a & \dfrac{\gamma}{\beta} f(x_\gamma, z_\beta) \\
b & (z \equiv x) \\
c & \vdash \dfrac{\gamma}{\beta} f(z_\gamma, x_\beta) \\
d & \dfrac{\gamma}{\beta} f(z_\gamma, x_\beta)
\end{array}
\qquad
\vdash\!\!\begin{array}{l}
\dfrac{\gamma}{\beta} f(x_\gamma, z_\beta) \\
\dfrac{\gamma}{\beta} f(z_\gamma, x_\beta) \\
\dfrac{\gamma}{\beta} f(z_\gamma, x_\beta) \\
(z \equiv x) \\
\dfrac{\gamma}{\beta} f(z_\gamma, x_\beta) \\
\dfrac{\gamma}{\beta} f(z_\gamma, x_\beta)
\end{array}
\qquad (113).
$$

(104)::

$$\begin{array}{c|c} x & z \\ z & x \end{array} \qquad \vdash \begin{array}{l} \underset{\beta}{\overset{\gamma}{\equiv}} f(x_\gamma, z_\beta) \\ \underset{\beta}{\overset{\gamma}{\equiv}} f(z_\gamma, x_\beta) \\ \underset{\beta}{\overset{\gamma}{\equiv}} f(z_\gamma, x_\beta) \end{array} \tag{114}$$

In words the derivation of this formula is as follows:

Assume that x belongs to the f-sequence beginning with z. Then by (104) z is the same as x or x follows z in the f-sequence. If z is the same as x, then by (112) z belongs to the f-sequence beginning with x. From the last two propositions it follows that z belongs to the f-sequence beginning with x or x follows z in the f-sequence. Therefore,

If x belongs to the f-sequence beginning with z, then z belongs to the f-sequence beginning with x or x follows z in the f-sequence.

§ 31.
$$\Vdash \left[\left[\begin{array}{l} (\mathfrak{a} \equiv \mathfrak{e}) \\ f(\mathfrak{b}, \mathfrak{a}) \\ f(\mathfrak{b}, \mathfrak{e}) \end{array} \right] \equiv \underset{\varepsilon}{\overset{\delta}{I}} f(\delta, \varepsilon) \right]^{24} \tag{115}$$

I translate
$$\underset{\varepsilon}{\overset{\delta}{I}} f(\delta, \varepsilon)$$

by "the circumstance that the procedure f is single-valued". Then (115) can be rendered thus:

If from the circumstance that $\mathfrak{e}$ is a result of an application of the procedure f to $\mathfrak{b}$, whatever $\mathfrak{b}$ may be, it can be inferred that every result of an application of the procedure f to $\mathfrak{b}$ is the same as $\mathfrak{e}$, then I say: "The procedure f is single-valued".

115
$$\vdash \left[\left[\begin{array}{l} (\mathfrak{a} \equiv \mathfrak{e}) \\ f(\mathfrak{b}, \mathfrak{a}) \\ f(\mathfrak{b}, \mathfrak{e}) \end{array} \right] \equiv \underset{\varepsilon}{\overset{\delta}{I}} f(\delta, \varepsilon) \right]$$

(68):
$$\begin{array}{c|l} f(\Gamma) & \begin{array}{l}(\mathfrak{a} \equiv \Gamma) \\ f(\mathfrak{b}, \mathfrak{a}) \\ f(\mathfrak{b}, \Gamma)\end{array} \qquad \vdash \begin{array}{l}(\mathfrak{a} \equiv x) \\ f(\mathfrak{b}, \mathfrak{a}) \\ f(\mathfrak{b}, x)\end{array} \\ \mathfrak{b} & \underset{\varepsilon}{\overset{\delta}{I}} f(\delta, \varepsilon) \qquad\qquad\qquad \underset{\varepsilon}{\overset{\delta}{I}} f(\delta, \varepsilon) \\ c & x \\ \mathfrak{a} & \mathfrak{e} \end{array} \tag{116}$$

(9):

[24] See § 24.

BEGRIFFSSCHRIFT

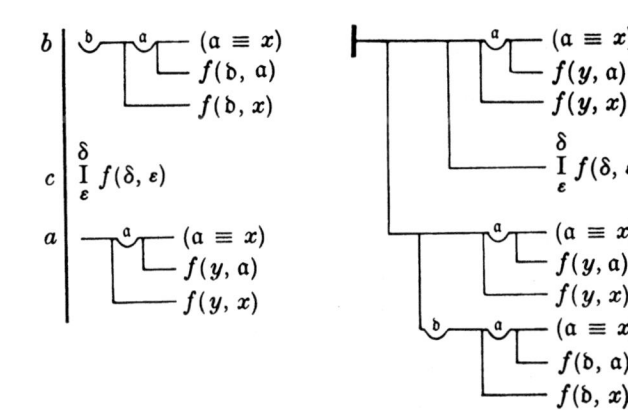

(117).

(58) ::

———

(118).

(19) :

———

(119).

(58) ::

———

(120).

(20) :

———

76   FREGE

$$
\begin{array}{c|l}
b & (\mathfrak{a} \equiv x) \\
c & f(y, \mathfrak{a}) \\
d & f(y, x) \\
e & \mathop{I}\limits_{\varepsilon}^{\delta} f(\delta, \varepsilon) \\
a & \dfrac{\gamma}{\bar{\beta}} f(x_\gamma, \mathfrak{a}_\beta)
\end{array}
$$

$$
\begin{array}{l}
\vdash\!\!\!\!\begin{array}{l}
\dfrac{\gamma}{\bar{\beta}} f(x_\gamma, \mathfrak{a}_\beta) \\
f(y, \mathfrak{a}) \\
f(y, x) \\
\mathop{I}\limits_{\varepsilon}^{\delta} f(\delta, \varepsilon) \\
\dfrac{\gamma}{\bar{\beta}} f(x_\gamma, \mathfrak{a}_\beta) \\
(\mathfrak{a} \equiv x)
\end{array}
\end{array}
\qquad (121).
$$

(112) ::

$z \mid \mathfrak{a}$

$$
\vdash\!\!\!\!\begin{array}{l}
\dfrac{\gamma}{\bar{\beta}} f(x_\gamma, \mathfrak{a}_\beta) \\
f(y, \mathfrak{a}) \\
f(y, x) \\
\mathop{I}\limits_{\varepsilon}^{\delta} f(\delta, \varepsilon)
\end{array}
\qquad (122).
$$

122
$\mathfrak{a} \mid \mathfrak{a}$

$$
\vdash\!\!\!\!\begin{array}{l}
\overset{\mathfrak{a}}{\smile}\dfrac{\gamma}{\bar{\beta}} f(x_\gamma, \mathfrak{a}_\beta) \\
f(y, \mathfrak{a}) \\
f(y, x) \\
\mathop{I}\limits_{\varepsilon}^{\delta} f(\delta, \varepsilon)
\end{array}
$$

(19) :

$$
\begin{array}{c|l}
b & \overset{\mathfrak{a}}{\smile} \dfrac{\gamma}{\bar{\beta}} f(x_\gamma, \mathfrak{a}_\beta) \\
  & f(y, \mathfrak{a}) \\
c & f(y, x) \\
d & \mathop{I}\limits_{\varepsilon}^{\delta} f(\delta, \varepsilon) \\
a & \begin{array}{l}\dfrac{\gamma}{\bar{\beta}} f(x_\gamma, m_\beta) \\ \dfrac{\gamma}{\bar{\beta}} f(y_\gamma, m_\beta)\end{array}
\end{array}
$$

$$
\vdash\!\!\!\!\begin{array}{l}
\dfrac{\gamma}{\bar{\beta}} f(x_\gamma, m_\beta) \\
\dfrac{\gamma}{\bar{\beta}} f(y_\gamma, m_\beta) \\
f(y, x) \\
\mathop{I}\limits_{\varepsilon}^{\delta} f(\delta, \varepsilon) \\
\dfrac{\gamma}{\bar{\beta}} f(x_\gamma, m_\beta) \\
\dfrac{\gamma}{\bar{\beta}} f(y_\gamma, m_\beta) \\
\overset{\mathfrak{a}}{\smile} \dfrac{\gamma}{\bar{\beta}} f(x_\gamma, \mathfrak{a}_\beta) \\
f(y, \mathfrak{a})
\end{array}
\qquad (123).
$$

(110) ::

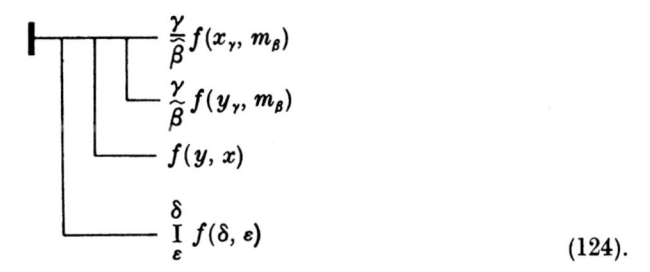

$$\vdash \begin{array}{l} \underset{\beta}{\overset{\gamma}{=}} f(x_\gamma, m_\beta) \\ \underset{\beta}{\overset{\gamma}{=}} f(y_\gamma, m_\beta) \\ f(y, x) \\ \underset{\varepsilon}{\overset{\delta}{\mathrm{I}}} f(\delta, \varepsilon) \end{array} \qquad (124).$$

Let us give the derivation of formulas (122) and (124) in words.

Assume that x is a result of an application of the single-valued procedure f to y. Then by (120) every result of an application of the procedure f to y is the same as x. Hence by (112) every result of an application of the procedure f to y belongs to the f-sequence beginning with x. Therefore,

If x is a result of an application of the single-valued procedure f to y, then every result of an application of the procedure f to y belongs to the f-sequence beginning with x. (Formula (122).)

Assume that m follows y in the f-sequence. Then (110) yields: If every result of an application of the procedure f to y belongs to the f-sequence beginning with x, then m belongs to the f-sequence beginning with x. This, combined with (122), shows that, if x is a result of an application of the single-valued procedure f to y, then m belongs to the f-sequence beginning with x. Therefore,

If x is a result of an application of the single-valued procedure f to y and if m follows y in the f-sequence, then m belongs to the f-sequence beginning with x. (Formula (124).)

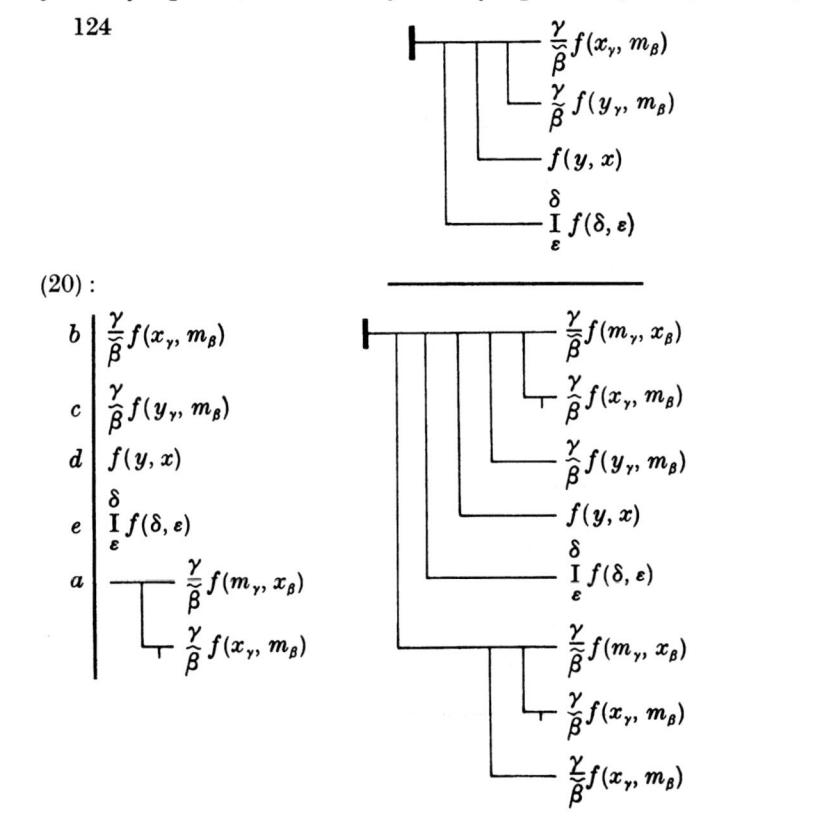

124

$$\vdash \begin{array}{l} \underset{\beta}{\overset{\gamma}{=}} f(x_\gamma, m_\beta) \\ \underset{\beta}{\overset{\gamma}{=}} f(y_\gamma, m_\beta) \\ f(y, x) \\ \underset{\varepsilon}{\overset{\delta}{\mathrm{I}}} f(\delta, \varepsilon) \end{array}$$

(20):

$$\begin{array}{rl} b & \underset{\beta}{\overset{\gamma}{=}} f(x_\gamma, m_\beta) \\ c & \underset{\beta}{\overset{\gamma}{=}} f(y_\gamma, m_\beta) \\ d & f(y, x) \\ e & \underset{\varepsilon}{\overset{\delta}{\mathrm{I}}} f(\delta, \varepsilon) \\ a & \underset{\beta}{\overset{\gamma}{=}} f(m_\gamma, x_\beta) \\ & \underset{\beta}{\overset{\gamma}{=}} f(x_\gamma, m_\beta) \end{array}$$

$$\vdash \begin{array}{l} \underset{\beta}{\overset{\gamma}{=}} f(m_\gamma, x_\beta) \\ \underset{\beta}{\overset{\gamma}{=}} f(x_\gamma, m_\beta) \\ \underset{\beta}{\overset{\gamma}{=}} f(y_\gamma, m_\beta) \\ f(y, x) \\ \underset{\varepsilon}{\overset{\delta}{\mathrm{I}}} f(\delta, \varepsilon) \\ \underset{\beta}{\overset{\gamma}{=}} f(m_\gamma, x_\beta) \\ \underset{\beta}{\overset{\gamma}{=}} f(x_\gamma, m_\beta) \\ \underset{\beta}{\overset{\gamma}{=}} f(x_\gamma, m_\beta) \end{array} \qquad (125).$$

(114) ::

$$\begin{array}{c|c} x & m \\ z & x \end{array}$$

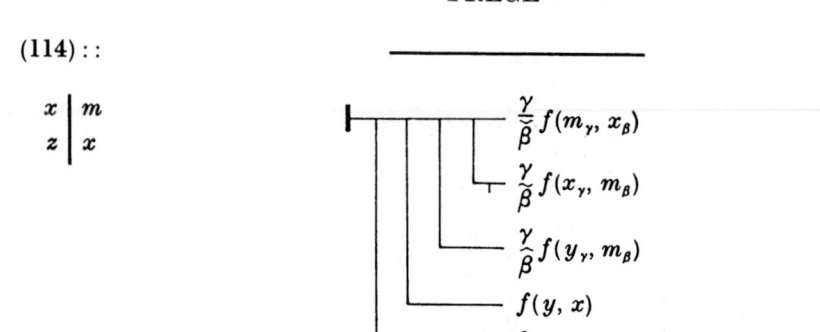

$$\vdash \begin{array}{l} \underset{\beta}{\overset{\gamma}{\approx}} f(m_\gamma, x_\beta) \\ \underset{\beta}{\overset{\gamma}{\approx}} f(x_\gamma, m_\beta) \\ \underset{\beta}{\overset{\gamma}{\approx}} f(y_\gamma, m_\beta) \\ f(y, x) \\ \underset{\epsilon}{\overset{\delta}{\text{I}}} f(\delta, \epsilon) \end{array}$$ (126).

The derivation of this formula follows here in words.

Assume that x is a result of an application of the single-valued procedure f to y. Assume that m follows y in the f-sequence. Then by (124) m belongs to the f-sequence beginning with x. Consequently, by (114) x belongs to the f-sequence beginning with m or m follows x in the f-sequence. This can also be expressed as follows: x belongs to the f-sequence beginning with m or precedes m in the f-sequence. Therefore,

If m follows y in the f-sequence and if the procedure f is single-valued, then every result of an application of the procedure f to y belongs to the f-sequence beginning with m or precedes m in the f-sequence.

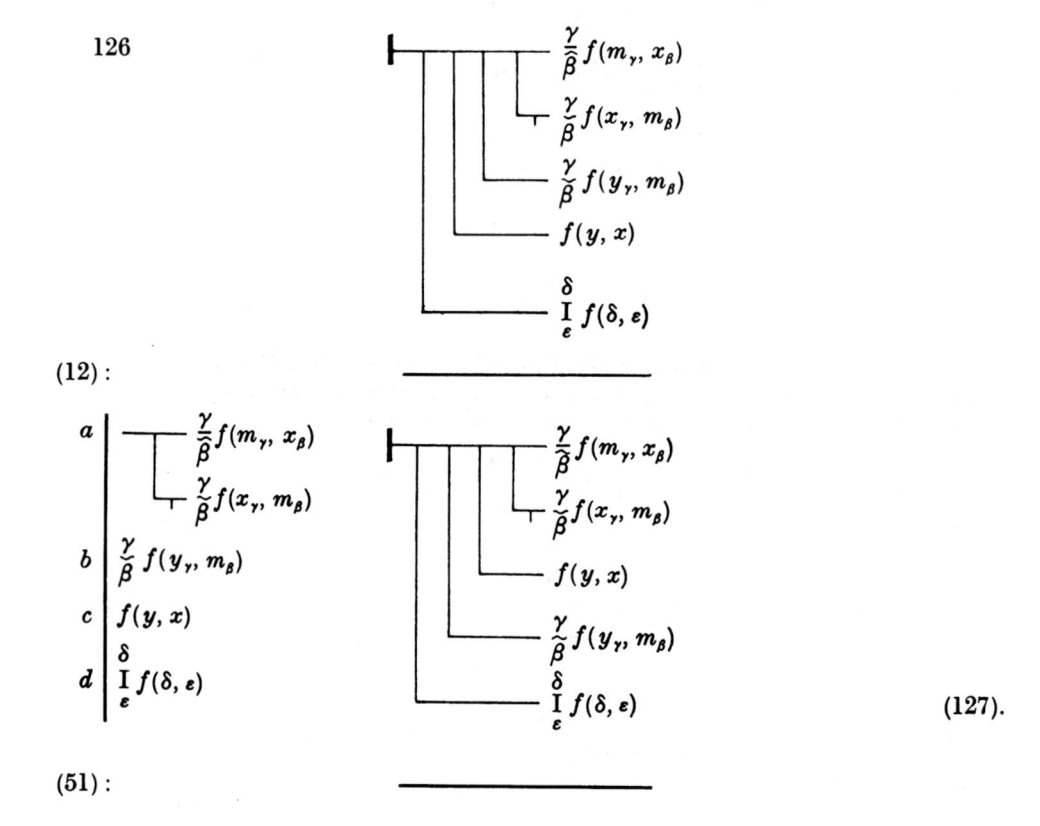

126

$$\vdash \begin{array}{l} \underset{\beta}{\overset{\gamma}{\approx}} f(m_\gamma, x_\beta) \\ \underset{\beta}{\overset{\gamma}{\approx}} f(x_\gamma, m_\beta) \\ \underset{\beta}{\overset{\gamma}{\approx}} f(y_\gamma, m_\beta) \\ f(y, x) \\ \underset{\epsilon}{\overset{\delta}{\text{I}}} f(\delta, \epsilon) \end{array}$$

(12) :

$$\begin{array}{c|l} a & \underset{\beta}{\overset{\gamma}{\approx}} f(m_\gamma, x_\beta) \\ & \underset{\beta}{\overset{\gamma}{\approx}} f(x_\gamma, m_\beta) \\ b & \underset{\beta}{\overset{\gamma}{\approx}} f(y_\gamma, m_\beta) \\ c & f(y, x) \\ d & \underset{\epsilon}{\overset{\delta}{\text{I}}} f(\delta, \epsilon) \end{array}$$

$$\vdash \begin{array}{l} \underset{\beta}{\overset{\gamma}{\approx}} f(m_\gamma, x_\beta) \\ \underset{\beta}{\overset{\gamma}{\approx}} f(x_\gamma, m_\beta) \\ f(y, x) \\ \underset{\beta}{\overset{\gamma}{\approx}} f(y_\gamma, m_\beta) \\ \underset{\epsilon}{\overset{\delta}{\text{I}}} f(\delta, \epsilon) \end{array}$$ (127).

(51) :

BEGRIFFSSCHRIFT 79

$$a \begin{cases} \vdash\!\!\!-\dfrac{\gamma}{\beta}f(m_\gamma, x_\beta) \\ \!\!\!-\!\!\top\dfrac{\gamma}{\beta}f(x_\gamma, m_\beta) \\ \!\!\!-f(y, x) \end{cases}$$

$c \mid \dfrac{\gamma}{\beta}f(y_\gamma, m_\beta)$

$d \mid \underset{\varepsilon}{\overset{\delta}{\mathrm{I}}} f(\delta, \varepsilon)$

$b \mid \dfrac{\gamma}{\beta}f(m_\gamma, y_\beta)$

$$\vdash \begin{cases} \dfrac{\gamma}{\beta}f(m_\gamma, x_\beta) \\ \top\dfrac{\gamma}{\beta}f(x_\gamma, m_\beta) \\ f(y, x) \\ \dfrac{\gamma}{\beta}f(m_\gamma, y_\beta) \\ \top\dfrac{\gamma}{\beta}f(y_\gamma, m_\beta) \\ \underset{\varepsilon}{\overset{\delta}{\mathrm{I}}} f(\delta, \varepsilon) \\ \dfrac{\gamma}{\beta}f(m_\gamma, x_\beta) \\ \top\dfrac{\gamma}{\beta}f(x_\gamma, m_\beta) \\ f(y, x) \\ \dfrac{\gamma}{\beta}f(m_\gamma, y_\beta) \end{cases}$$ (128).

(111)::

$\begin{array}{c|c} z & m \\ v & x \end{array}$

$$\vdash \begin{cases} \dfrac{\gamma}{\beta}f(m_\gamma, x_\beta) \\ \top\dfrac{\gamma}{\beta}f(x_\gamma, m_\beta) \\ f(y, x) \\ \dfrac{\gamma}{\beta}f(m_\gamma, y_\beta) \\ \top\dfrac{\gamma}{\beta}f(y_\gamma, m_\beta) \\ \underset{\varepsilon}{\overset{\delta}{\mathrm{I}}} f(\delta, \varepsilon) \end{cases}$$ (129).

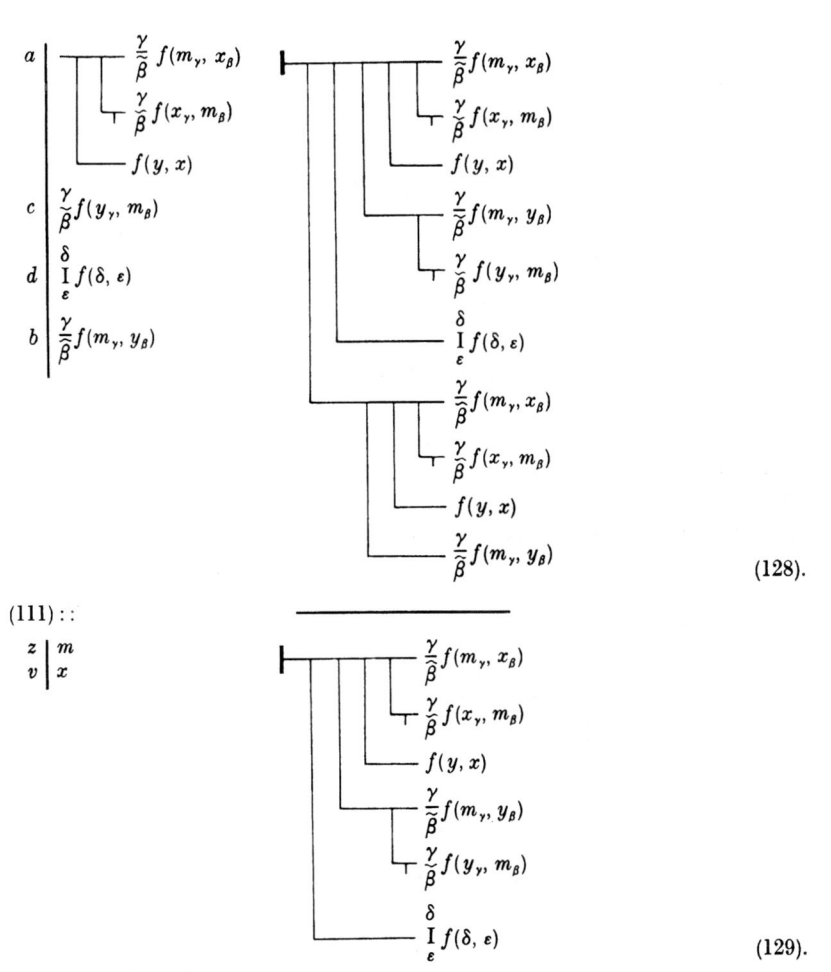

In words (129) reads:

If the procedure f is single-valued and y belongs to the f-sequence beginning with m or precedes m in the f-sequence, then every result of an application of the procedure f to y belongs to the f-sequence beginning with m or precedes m in the f-sequence.

129

$\begin{array}{c|c} x & \mathfrak{a} \\ y & \mathfrak{b} \end{array}$

$$\vdash_{\mathfrak{b}}{}^{\mathfrak{a}} \begin{cases} \dfrac{\gamma}{\beta}f(m_\gamma, \mathfrak{a}_\beta) \\ \top\dfrac{\gamma}{\beta}f(\mathfrak{a}_\gamma, m_\beta) \\ f(\mathfrak{b}, \mathfrak{a}) \\ \dfrac{\gamma}{\beta}f(m_\gamma, \mathfrak{b}_\beta) \\ \top\dfrac{\gamma}{\beta}f(\mathfrak{b}_\gamma, m_\beta) \\ \underset{\varepsilon}{\overset{\delta}{\mathrm{I}}} f(\delta, \varepsilon) \end{cases}$$

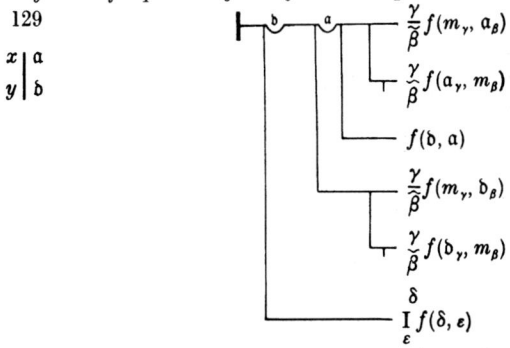

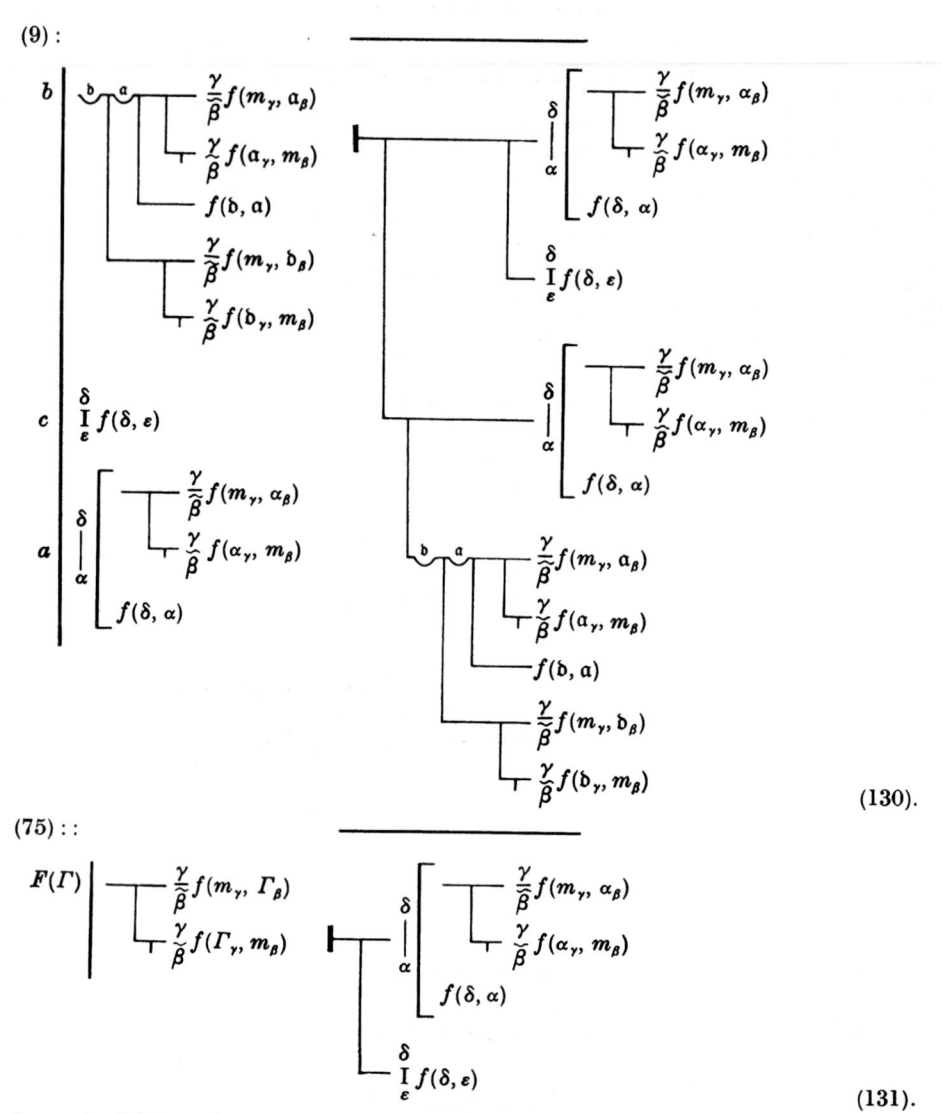

(9):
$$b \left\{ \underset{a}{\overset{b}{\frown}} \begin{array}{l} \underset{\beta}{\overset{\gamma}{=}} f(m_\gamma, a_\beta) \\ \underset{\beta}{\overset{\gamma}{=}} f(a_\gamma, m_\beta) \\ f(b, a) \\ \underset{\beta}{\overset{\gamma}{=}} f(m_\gamma, b_\beta) \\ \underset{\beta}{\overset{\gamma}{=}} f(b_\gamma, m_\beta) \end{array} \right.$$

$c \;\; \underset{\varepsilon}{\overset{\delta}{I}} f(\delta, \varepsilon)$

$a \;\; \underset{\alpha}{\overset{\delta}{|}} \left[\begin{array}{l} \underset{\beta}{\overset{\gamma}{=}} f(m_\gamma, \alpha_\beta) \\ \underset{\beta}{\overset{\gamma}{=}} f(\alpha_\gamma, m_\beta) \\ f(\delta, \alpha) \end{array} \right.$

$$\vdash \underset{\alpha}{\overset{\delta}{|}} \left[\begin{array}{l} \underset{\beta}{\overset{\gamma}{=}} f(m_\gamma, \alpha_\beta) \\ \underset{\beta}{\overset{\gamma}{=}} f(\alpha_\gamma, m_\beta) \\ f(\delta, \alpha) \end{array} \right.$$
$$\underset{\varepsilon}{\overset{\delta}{I}} f(\delta, \varepsilon)$$
$$\underset{\alpha}{\overset{\delta}{|}} \left[\begin{array}{l} \underset{\beta}{\overset{\gamma}{=}} f(m_\gamma, \alpha_\beta) \\ \underset{\beta}{\overset{\gamma}{=}} f(\alpha_\gamma, m_\beta) \\ f(\delta, \alpha) \end{array} \right.$$
$$\underset{a}{\overset{b}{\frown}} \begin{array}{l} \underset{\beta}{\overset{\gamma}{=}} f(m_\gamma, a_\beta) \\ \underset{\beta}{\overset{\gamma}{=}} f(a_\gamma, m_\beta) \\ f(b, a) \\ \underset{\beta}{\overset{\gamma}{=}} f(m_\gamma, b_\beta) \\ \underset{\beta}{\overset{\gamma}{=}} f(b_\gamma, m_\beta) \end{array}$$ (130).

(75)::

$F(\Gamma) \left| \begin{array}{l} \underset{\beta}{\overset{\gamma}{=}} f(m_\gamma, \Gamma_\beta) \\ \underset{\beta}{\overset{\gamma}{=}} f(\Gamma_\gamma, m_\beta) \end{array} \right. \vdash \begin{array}{l} \underset{\alpha}{\overset{\delta}{|}} \left[\begin{array}{l} \underset{\beta}{\overset{\gamma}{=}} f(m_\gamma, \alpha_\beta) \\ \underset{\beta}{\overset{\gamma}{=}} f(\alpha_\gamma, m_\beta) \\ f(\delta, \alpha) \end{array} \right. \\ \underset{\varepsilon}{\overset{\delta}{I}} f(\delta, \varepsilon) \end{array}$ (131).

In words (131) reads:

If the procedure f is single-valued, then the property of belonging to the f-sequence beginning with m or of preceding m in the f-sequence is hereditary in the f-sequence.

131

$$\vdash \underset{\alpha}{\overset{\delta}{|}} \left[\begin{array}{l} \underset{\beta}{\overset{\gamma}{=}} f(m_\gamma, \alpha_\beta) \\ \underset{\beta}{\overset{\gamma}{=}} f(\alpha_\gamma, m_\beta) \\ f(\delta, \alpha) \end{array} \right.$$
$$\underset{\varepsilon}{\overset{\delta}{I}} f(\delta, \varepsilon)$$

BEGRIFFSSCHRIFT

(9):

$$b \left| \begin{array}{l} \delta \\ | \\ \alpha \end{array} \right. \left[\begin{array}{l} \dfrac{\gamma}{\check{\beta}} f(m_\gamma, \alpha_\beta) \\ \dfrac{\gamma}{\check{\beta}} f(\alpha_\gamma, m_\beta) \\ f(\delta, \alpha) \end{array} \right.$$

$$c \left| \begin{array}{l} \delta \\ \mathrm{I} \\ \varepsilon \end{array} f(\delta, \varepsilon) \right.$$

$$a \left| \begin{array}{l} \dfrac{\gamma}{\check{\beta}} f(m_\gamma, y_\beta) \\ \dfrac{\gamma}{\check{\beta}} f(y_\gamma, m_\beta) \\ \dfrac{\gamma}{\check{\beta}} f(x_\gamma, y_\beta) \\ \dfrac{\gamma}{\check{\beta}} f(x_\gamma, m_\beta) \end{array} \right.$$

$$\begin{array}{l} \dfrac{\gamma}{\check{\beta}} f(m_\gamma, y_\beta) \\ \dfrac{\gamma}{\check{\beta}} f(y_\gamma, m_\beta) \\ \dfrac{\gamma}{\check{\beta}} f(x_\gamma, y_\beta) \\ \dfrac{\gamma}{\check{\beta}} f(x_\gamma, m_\beta) \\ \mathrm{I}_\varepsilon^\delta f(\delta, \varepsilon) \\ \dfrac{\gamma}{\check{\beta}} f(m_\gamma, y_\beta) \\ \dfrac{\gamma}{\check{\beta}} f(y_\gamma, m_\beta) \\ \dfrac{\gamma}{\check{\beta}} f(x_\gamma, y_\beta) \\ \dfrac{\gamma}{\check{\beta}} f(x_\gamma, m_\beta) \\ \begin{array}{l} \delta \\ | \\ \alpha \end{array} \left[\begin{array}{l} \dfrac{\gamma}{\check{\beta}} f(m_\gamma, \alpha_\beta) \\ \dfrac{\gamma}{\check{\beta}} f(\alpha_\gamma, m_\beta) \\ f(\delta, \alpha) \end{array} \right. \end{array}$$

(132).

(83)::

$$\begin{array}{l} g(\Gamma) \\ h(\Gamma) \end{array} \left| \begin{array}{l} \dfrac{\gamma}{\check{\beta}} f(m_\gamma, \Gamma_\beta) \\ \dfrac{\gamma}{\check{\beta}} f(\Gamma_\gamma, m_\beta) \end{array} \right.$$

$$\begin{array}{l} \dfrac{\gamma}{\check{\beta}} f(m_\gamma, y_\beta) \\ \dfrac{\gamma}{\check{\beta}} f(y_\gamma, m_\beta) \\ \dfrac{\gamma}{\check{\beta}} f(x_\gamma, y_\beta) \\ \dfrac{\gamma}{\check{\beta}} f(x_\gamma, m_\beta) \\ \mathrm{I}_\varepsilon^\delta f(\delta, \varepsilon) \end{array}$$

(133).

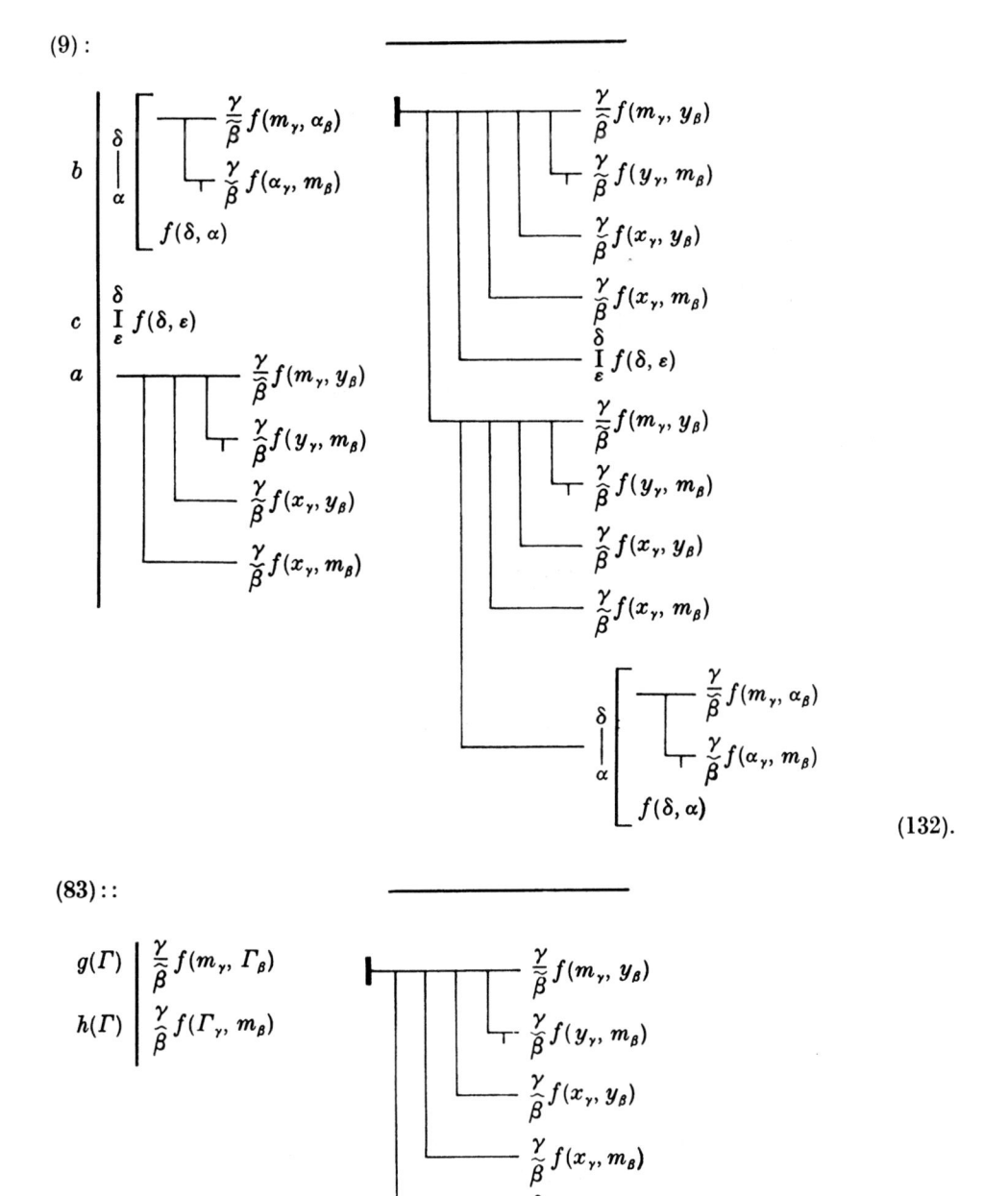

In words this proposition reads:

If the procedure f is single-valued and if m and y follow x in the f-sequence, then y belongs to the f-sequence beginning with m or precedes m in the f-sequence.

82   FREGE

Below I give a table that shows where use has been made of one formula in the derivation of another. The table can be used to look up the ways in which a formula has been employed. From it we can also see how frequently a formula has been used.

The right column always contains the number of the formula in whose derivation the one listed in the left column was used.

1	3	7	94	12	35	23	48	47	49	63	91	84	98	109	110
1	5	7	107	12	49	24	25	48	101	64	65	85	86	110	124
1	11	7	113	12	60	24	63	49	50	65	66	86	87	111	129
1	24	8	9	12	85	25	111	50	51	66	—	87	88	112	113
1	26	8	10	12	127	26	27	51	128	67	68	88	95	112	122
1	27	8	12	13	14	27	42	52	53	68	70	89	90	113	114
1	36	8	17	14	15	28	29	52	57	68	77	90	91	114	126
2	3	8	26	15	88	28	33	52	75	68	116	90	93	115	116
2	4	8	38	16	17	29	30	52	89	69	70	91	92	116	117
2	39	8	53	16	18	30	59	52	105	69	75	92	102	117	118
2	73	8	62	16	22	31	32	53	55	70	71	93	94	118	119
2	79	8	66	17	50	32	33	53	92	71	72	94	95	119	120
3	4	8	74	17	78	33	34	54	55	72	73	95	96	120	121
4	5	8	84	18	19	33	46	55	56	72	74	96	97	121	122
5	6	8	96	18	20	34	35	55	104	73	87	96	102	122	123
5	7	9	10	18	23	34	36	56	57	74	81	97	98	123	124
5	9	9	11	18	51	35	40	57	68	75	97	98	—	124	125
5	12	9	19	18	64	36	37	57	100	75	109	99	100	125	126
5	14	9	21	18	82	36	38	58	59	75	131	99	105	126	127
5	16	9	37	19	20	36	83	58	60	76	77	100	101	127	128
5	18	9	56	19	21	37	106	58	61	76	89	100	103	128	129
5	22	9	61	19	71	38	39	58	62	77	78	101	102	129	130
5	25	9	117	19	86	39	40	58	67	77	85	102	108	130	131
5	29	9	130	19	103	40	43	58	72	78	79	103	104	131	132
5	34	9	132	19	119	41	42	58	118	78	110	104	114	132	133
5	45	10	30	19	123	42	43	58	120	79	80	105	106	133	—
5	80	11	112	20	121	43	44	59	—	80	81	105	112		
5	90	12	13	20	125	44	45	60	93	81	82	106	107		
6	7	12	15	21	44	45	46	61	65	81	84	107	108		
7	32	12	16	21	47	46	47	62	63	82	83	108	109		
7	67	12	24	22	23	47	48	62	64	83	133	108	111		

Some metamathematical results on completeness and consistency,
On formally undecidable propositions of
Principia mathematica *and related systems I,*
and
On completeness and consistency

KURT GÖDEL

(*1930b*, *1931*, and *1931a*)

The main paper below (*1931*), which was to have such an impact on modern logic, was received for publication on 17 November 1930 and published early in 1931. An abstract (*1930b*) had been presented on 23 October 1930 to the Vienna Academy of Sciences by Hans Hahn.

Gödel's results are now accessible in many publications, but his original paper has not lost any of its value as a guide. It is clearly written and does not assume any previous result for its main line of argument. It is, moreover, rich in interesting details. We now give some indications of its contents and structure.

Section 1 is an informal presentation of the main argument and can be read by the nonmathematician; it shows how the argument, by dealing with the proposition that states of itself "I am not provable", instead of the proposition that states of itself "I am not true", skirts the Liar paradox, without falling into it. Gödel also brings to light the relation that his argument bears to Cantor's diagonal procedure and to Richard's paradox (Herbrand (*1931b*, pp. 7–8) and Weyl (*1949*, pp. 219–235) lay particular stress on this aspect of Gödel's argument).

Section 2, the longest, is the proof of Theorem VI. The theorem states that in a formal system satisfying certain precise conditions there is an undecidable proposition, that is, a proposition such that neither the proposition itself nor its negation is provable in the system. Before coming to the core of the argument, Gödel takes a number of preparatory steps:

(1) A precise description of the system P with which he is going to work. The variables are distinguished as to their types and they range over the natural numbers (type 1), classes of natural numbers (type 2), classes of classes of natural numbers (type 3), and so forth. The logical axioms are equivalent to the logic of *Principia mathematica* without the ramified theory of types. The arithmetic axioms are Peano's, properly transcribed. The identification of the individuals with the natural numbers and

the adjunction of Peano's axioms (instead of their derivation, as in *Principia*) have the effect that every formula has an interpretation in classical mathematics and, if closed, is either true or false in that interpretation; moreover, proofs are considerably shortened.

(2) An assignment of natural numbers to sequences of signs of P and a similar assignment to sequences of sequences of signs of P. The first assignment is such that, given a sequence, the number assigned to it can be effectively calculated, and, given a number, we can effectively decide whether the number is assigned to a sequence and, if it is, actually write down the sequence; similarly for the second assignment. By means of these assignments we can correlate number-theoretic predicates with metamathematical notions used in the description of the system; for example, to the notion "axiom" corresponds the predicate $Ax(x)$, which holds precisely of the numbers x that are assigned to axioms (the "Gödel numbers" of axioms, we would say today).

(3) A definition of primitive recursive functions (Gödel calls them recursive functions) and the derivation of a few theorems about them. These functions had already been used in foundational research (for example, by Dedekind (*1888*), Skolem (*1923*), Hilbert (*1925, 1927*), and Ackermann (*1928*)); Gödel gives a precise definition of them, which has become standard.

(4) The proof that forty-five number-theoretic predicates, forty of them associated with metamathematical notions, are primitive recursive.

(5) The proof that every primitive recursive number-theoretic predicate is numeralwise representable in P. That is, the predicate holds of some given numbers if and only if a definite formula of P is provable whenever its free variables are replaced by the symbols that represent these numbers in P.

(6) The definition of ω-consistency.

Gödel can then undertake to prove Theorem VI. The scope of the theorem is enlarged by the addition of any ω-consistent primitive recursive class κ of formulas to the axioms of P. For each such κ a different system is thus obtained (in the present note, "P_κ", a notation not used by Gödel, will denote the system corresponding to a given κ). After the proof Gödel makes a number of important remarks:

(*a*) He points out the constructive content of Theorem VI.

(*b*) He introduces predicates that are *entscheidungsdefinit* (in the translation below these are called *decidable* predicates, at the author's suggestion). If we take into account the few lines added in proof at the end of a later note of Gödel's (*1934a*), these predicates are in fact those that today we call recursive (that is, general recursive) predicates. Gödel somewhat extends the result of Theorem VI by assuming only that κ is decidable, and not that it is primitive recursive.

(*c*) If κ is assumed to be merely consistent, instead of ω-consistent, the proof yields the existence of a predicate whose universalization is not provable but for which no counterexample can be given; P_κ is ω-incomplete, as we would say today.

(*d*) The adjunction of the undecidable formula Neg(17 Gen r) to κ yields a consistent but ω-inconsistent system.

(*e*) Even with the adjunction of the axiom of choice or the continuum hypothesis the system contains undecidable propositions.

The section ends with a review of the properties of P that are actually used in the proof and the remark that all known axiom systems of mathematics, or of any substantial part of it, have these properties.

Section 3 presents two supplementary undecidability results. Gödel establishes (Theorem VII) that a primitive recursive number-theoretic predicate is *arithmetic*, that is, can be expressed as a formula of first-order number theory (this yields a stronger result than the numeralwise representability of such predicates, as it was introduced and used in Section 2). Hence every formula of the form $(x)F(x)$, with $F(x)$ primitive recursive, is equivalent to an arithmetic formula; moreover, this equivalence is provable in P_κ: one can review the informal proof presented by Gödel and check that P_κ is strong enough to express and justify each of its steps. Since the proposition that was proved to be undecidable in Theorem VI is of the form $(x)F(x)$, with $F(x)$ primitive recursive, P_κ contains undecidable *arithmetic* propositions (Theorem VIII). For all its strength, the system P_κ cannot decide every first-order number-theoretic proposition. Theorem X states that, given a formula $(x)F(x)$, with $F(x)$ primitive recursive, one can exhibit a formula of the *pure* first-order predicate calculus, say A, that is satisfiable if and only if $(x)F(x)$ holds. Moreover, since P_κ contains a set theory, the equivalence

$$(x)F(x) \equiv (A \text{ is satisfiable})$$

is expressible in P_κ and, as one can verify by reviewing Gödel's informal argument, provable in P_κ. Therefore (Theorem IX) there are formulas of the pure first-order predicate calculus whose validity is undecidable in P_κ.

In Section 4 an important consequence of Theorem VI is derived. The statement "there exists in P_κ an unprovable formula", which expresses the consistency of P_κ, can be written as a formula of P_κ; but this formula is not provable in P_κ (Theorem XI). The main step in the demonstration of this result consists in reviewing the proof of the first half of Theorem VI and checking that all the statements made in that proof can be expressed and proved in P_κ. It is clear that this is the case, and Gödel does not go through the details of the demonstration. The section ends with various remarks on Theorem XI (its constructive character, its applicability to set theory and ordinary analysis, its effect upon Hilbert's conception of mathematics).

Gödel's paper immediately attracted the interest of logicians and, although it caused some momentary surprise, its results were soon widely accepted. A number of studies were directly inspired by it. By using a somewhat more complicated predicate than "is provable in P", Rosser (*1936*) was able to weaken the assumption of ω-consistency in Theorem VI to that of ordinary consistency. Hilbert and Bernays (*1939*, pp. 283–340) carried out in all details the proof of the analogue of Theorem XI for two standard systems of number theory, Z_μ and Z, and this proof can be transferred almost literally to any system containing Z. As Gödel indicates in a note appended to the present translation of his paper, Turing's work (*1937*) gave to the notion of formal system its full generality. The notes of Gödel's Princeton lectures (*1934*) contain the most important results of the present paper, in a more succinct form; they also make precise the notion of (general) recursive function, already suggested by Herbrand (see *1931b*, p. 5). In developing the theory of these functions, Kleene (*1936*) obtained undecidability results of a somewhat different character from those presented here. Gödel's work led to Church's negative solution (*1936*) of the decision problem for the predicate calculus of first order. Tarski (*1953*) developed a general theory of undecidability. The device of the "arithmetization" of metamathematics became an everyday tool of the research worker in foundations. Gödel's results, finally, led to a profound revision of Hilbert's program (on that point see, among other texts, *Bernays 1938, 1954* and *Gödel 1958*).

These indications are far from giving a full account of the deep influence exerted in the field of foundations of mathematics by the results presented in the paper below and the methods used to obtain them. There is not one branch of research, except perhaps intuitionism, that has not been pervaded by this influence.

The translation of the paper is by the editor, and it is printed here with the kind permission of Professor Gödel and Springer Verlag. Professor Gödel approved the translation, which in many places was accommodated to his wishes. He suggested, in particular, the various phrases used to render the word "inhaltlich". He also proposed a number of short interpolations to help the reader, and these have been introduced in the text below between square brackets.

Below, on page 92, the author shows how a number-theoretic predicate can be associated with a given metamathematical notion and then used to represent the notion. In the German text such a predicate is denoted by the same word as the original notion, except that the word is printed in italics. Since in English italics are used for emphasis (while the German text uses letter spacing for that purpose), the translation below uses SMALL CAPITALS for the names of these predicates. This scheme of italicization (or small-capitalization), however, is used for only some of the number-theoretic predicates in question. According to Professor Gödel, "the idea was to use the notation only for those metamathematical notions that had been defined in their usual sense before, namely, those defined on pp. 90–92. From p. 98 up to the general considerations at the end of Section 2, and again in Section 4, every metamathematical term referring to the system P is supposed to denote the corresponding arithmetic one. But, of course, because of the complete isomorphism the distinction in many cases is entirely irrelevant".

Before the main text the reader will find a translation, by Stefan Bauer-Mengelberg, of its abstract (*1930b*); in that translation, at the author's suggestion, "entscheidungsdefinit", when referring to an axiom system, has been translated by "complete", and "Entscheidungsdefinitheit" by "completeness". A translation, by the editor, of *1931a*, a note dated 22 January 1931 and closely connected with *1931*, follows the main text. Both translations are printed here with the kind permission of Professor Gödel.

SOME METAMATHEMATICAL RESULTS ON COMPLETENESS AND CONSISTENCY

(*1930b*)

If to the Peano axioms we add the logic of *Principia mathematica*[1] (with the natural numbers as the individuals) together with the axiom of choice (for all types), we obtain a formal system S, for which the following theorems hold:

I. The system S is *not* complete [[entscheidungsdefinit]]; that is, it contains propositions A (and we can in fact exhibit such propositions) for which neither A nor $\bar{A}$ is provable and, in particular, it contains (even for decidable properties F of natural numbers) undecidable problems of the simple structure $(Ex)F(x)$, where x ranges over the natural numbers.[2]

II. Even if we admit all the logical devices of *Principia mathematica* (hence in particular the extended functional calculus[1] and the axiom of choice) in metamathematics, there does *not* exist a *consistency proof* for the system S (still less so if we

---

[1] With the axiom of reducibility or without ramified theory of types.

[2] Furthermore, S contains formulas of the restricted functional calculus such that neither universal validity nor existence of a counterexample is provable for any of them.

restrict the means of proof in any way). Hence a consistency proof for the system S can be carried out only by means of modes of inference that are not formalized in the system S itself, and analogous results hold for other formal systems as well, such as the Zermelo-Fraenkel axiom system of set theory.[3]

III. Theorem I can be sharpened to the effect that, even if we add finitely many axioms to the system S (or infinitely many that result from a finite number of them by "type elevation"), we do *not* obtain a complete system, provided the extended system is ω-consistent. Here a system is said to be ω-consistent if, for no property $F(x)$ of natural numbers,

$$F(1), F(2), \ldots, F(n), \ldots \text{ ad infinitum}$$

as well as

$$(Ex)\overline{F(x)}$$

are provable. (There are extensions of the system S that, while consistent, are not ω-consistent.)

IV. Theorem I still holds for all ω-consistent extensions of the system S that are obtained by the addition of *infinitely many* axioms, provided the added class of axioms is decidable [[entscheidungsdefinit]], that is, provided it is metamathematically decidable [[entscheidbar]] for every formula whether it is an axiom or not (here again we suppose that the logic used in metamathematics is that of *Principia mathematica*).

Theorems I, III, and IV can be extended also to other formal systems, for example, to the Zermelo-Fraenkel axiom system of set theory, provided the systems in question are ω-consistent.

The proofs of these theorems will appear in *Monatshefte für Mathematik und Physik*.

[3] This result, in particular, holds also for the axiom system of classical mathematics, as it has been constructed, for example, by von Neumann (*1927*).

ON FORMALLY UNDECIDABLE PROPOSITIONS OF *PRINCIPIA MATHEMATICA* AND RELATED SYSTEMS I[1]
(1931)

1

The development of mathematics toward greater precision has led, as is well known, to the formalization of large tracts of it, so that one can prove any theorem using nothing but a few mechanical rules. The most comprehensive formal systems that have been set up hitherto are the system of *Principia mathematica* (PM)[2] on the one hand and the Zermelo-Fraenkel axiom system of set theory (further developed by J. von Neumann)[3] on the other. These two systems are so comprehensive that in

[1] See a summary of the results of the present paper in *Gödel 1930b*.

[2] *Whitehead and Russell 1925*. Among the axioms of the system PM we include also the axiom of infinity (in this version: there are exactly denumerably many individuals), the axiom of reducibility, and the axiom of choice (for all types).

[3] See *Fraenkel 1927* and *von Neumann 1925, 1928*, and *1929*. We note that in order to complete the formalization we must add the axioms and rules of inference of the calculus of logic to the set-theoretic axioms given in the literature cited. The considerations that follow apply also to the formal systems (so far as they are available at present) constructed in recent years by Hilbert and his collaborators. See *Hilbert 1922, 1922a, 1927, Bernays 1923, von Neumann 1927*, and *Ackermann 1924*.

them all methods of proof today used in mathematics are formalized, that is, reduced to a few axioms and rules of inference. One might therefore conjecture that these axioms and rules of inference are sufficient to decide *any* mathematical question that can at all be formally expressed in these systems. It will be shown below that this is not the case, that on the contrary there are in the two systems mentioned relatively simple problems in the theory of integers[4] that cannot be decided on the basis of the axioms. This situation is not in any way due to the special nature of the systems that have been set up but holds for a wide class of formal systems; among these, in particular, are all systems that result from the two just mentioned through the addition of a finite number of axioms,[5] provided no false propositions of the kind specified in footnote 4 become provable owing to the added axioms.

Before going into details, we shall first sketch the main idea of the proof, of course without any claim to complete precision. The formulas of a formal system (we restrict ourselves here to the system *PM*) in outward appearance are finite sequences of primitive signs (variables, logical constants, and parentheses or punctuation dots), and it is easy to state with complete precision *which* sequences of primitive signs are meaningful formulas and which are not.[6] Similarly, proofs, from a formal point of view, are nothing but finite sequences of formulas (with certain specifiable properties.) Of course, for metamathematical considerations it does not matter what objects are chosen as primitive signs, and we shall assign natural numbers to this use.[7] Consequently, a formula will be a finite sequence of natural numbers,[8] and a proof array a finite sequence of finite sequences of natural numbers. The metamathematical notions (propositions) thus become notions (propositions) about natural numbers or sequences of them;[9] therefore they can (at least in part) be expressed by the symbols of the system *PM* itself. In particular, it can be shown that the notions "formula", "proof array", and "provable formula" can be defined in the system *PM*; that is, we can, for example, find a formula $F(v)$ of *PM* with one free variable v (of the type of a number sequence)[10] such that $F(v)$, interpreted according to the meaning of the terms of *PM*, says: v is a provable formula. We now construct an undecidable proposition of the system *PM*, that is, a proposition A for which neither A nor *not-A* is provable, in the following manner.

[4] That is, more precisely, there are undecidable propositions in which, besides the logical constants $-$ (not), $\vee$ (or), (x) (for all), and $=$ (identical with), no other notions occur but $+$ (addition) and $.$ (multiplication), both for natural numbers, and in which the prefixes (x), too, apply to natural numbers only.

[5] In *PM* only axioms that do not result from one another by mere change of type are counted as distinct.

[6] Here and in what follows we always understand by "formula of *PM*" a formula written without abbreviations (that is, without the use of definitions). It is well known that [in *PM*] definitions serve only to abbreviate notations and therefore are dispensable in principle.

[7] That is, we map the primitive signs one-to-one onto some natural numbers. (See how this is done on page 601.)

[8] That is, a number-theoretic function defined on an initial segment of the natural numbers. (Numbers, of course, cannot be arranged in a spatial order.)

[9] In other words, the procedure described above yields an isomorphic image of the system *PM* in the domain of arithmetic, and all metamathematical arguments can just as well be carried out in this isomorphic image. This is what we do below when we sketch the proof; that is, by "formula", "proposition", "variable", and so on, *we must always understand the corresponding objects of the isomorphic image*.

[10] It would be very easy (although somewhat cumbersome) to actually write down this formula.

A formula of *PM* with exactly one free variable, that variable being of the type of the natural numbers (class of classes), will be called a *class sign*. We assume that the class signs have been arranged in a sequence in some way,[11] we denote the *n*th one by $R(n)$, and we observe that the notion "class sign", as well as the ordering relation R, can be defined in the system *PM*. Let α be any class sign; by $[\alpha; n]$ we denote the formula that results from the class sign α when the free variable is replaced by the sign denoting the natural number n. The ternary relation $x = [y; z]$, too, is seen to be definable in *PM*. We now define a class K of natural numbers in the following way:

$$n \; \varepsilon \; K \equiv \overline{Bew} \, [R(n); n] \tag{1}$$

(where *Bew x* means: *x* is a provable formula).[11a] Since the notions that occur in the definiens can all be defined in *PM*, so can the notion K formed from them; that is, there is a class sign S such that the formula $[S; n]$, interpreted according to the meaning of the terms of *PM*, states that the natural number n belongs to K.[12] Since S is a class sign, it is identical with some $R(q)$; that is, we have

$$S = R(q)$$

for a certain natural number q. We now show that the proposition $[R(q); q]$ is undecidable in *PM*.[13] For let us suppose that the proposition $[R(q); q]$ were provable; then it would also be true. But in that case, according to the definitions given above, q would belong to K, that is, by (1), $\overline{Bew} \, [R(q); q]$ would hold, which contradicts the assumption. If, on the other hand, the negation of $[R(q); q]$ were provable, then $\overline{q \; \varepsilon \; K}$,[13a] that is, $Bew \, [R(q); q]$, would hold. But then $[R(q); q]$, as well as its negation, would be provable, which again is impossible.

The analogy of this argument with the Richard antinomy leaps to the eye. It is closely related to the "Liar" too;[14] for the undecidable proposition $[R(q); q]$ states that q belongs to K, that is, by (1), that $[R(q); q]$ is not provable. We therefore have before us a proposition that says about itself that it is not provable [in *PM*].[15] The method of proof just explained can clearly be applied to any formal system that, first, when interpreted as representing a system of notions and propositions, has at

---

[11] For example, by increasing sum of the finite sequence of integers that is the "class sign", and lexicographically for equal sums.

[11a] The bar denotes negation.

[12] Again, there is not the slightest difficulty in actually writing down the formula S.

[13] Note that "$[R(q); q]$" (or, which means the same, "$[S; q]$") is merely a *metamathematical description* of the undecidable proposition. But, as soon as the formula S has been obtained, we can, of course, also determine the number q and, therewith, actually write down the undecidable proposition itself. [This makes no difficulty in principle. However, in order not to run into formulas of entirely unmanageable lengths and to avoid practical difficulties in the computation of the number q, the construction of the undecidable proposition would have to be slightly modified, unless the technique of abbreviation by definition used throughout in *PM* is adopted.]

[13a] [[The German text reads $\overline{n \; \varepsilon \; K}$, which is a misprint.]]

[14] Any epistemological antinomy could be used for a similar proof of the existence of undecidable propositions.

[15] Contrary to appearances, such a proposition involves no faulty circularity, for initially it [only] asserts that a certain well-defined formula (namely, the one obtained from the *q*th formula in the lexicographic order by a certain substitution) is unprovable. Only subsequently (and so to speak by chance) does it turn out that this formula is precisely the one by which the proposition itself was expressed.

its disposal sufficient means of expression to define the notions occurring in the argument above (in particular, the notion "provable formula") and in which, second, every provable formula is true in the interpretation considered. The purpose of carrying out the above proof with full precision in what follows is, among other things, to replace the second of the assumptions just mentioned by a purely formal and much weaker one.

From the remark that $[R(q); q]$ says about itself that it is not provable it follows at once that $[R(q); q]$ is true, for $[R(q); q]$ *is* indeed unprovable (being undecidable). Thus, the proposition that is undecidable *in the system PM* still was decided by metamathematical considerations. The precise analysis of this curious situation leads to surprising results concerning consistency proofs for formal systems, results that will be discussed in more detail in Section 4 (Theorem XI).

2

We now proceed to carry out with full precision the proof sketched above. First we give a precise description of the formal system P for which we intend to prove the existence of undecidable propositions. P is essentially the system obtained when the logic of PM is superposed upon the Peano axioms[16] (with the numbers as individuals and the successor relation as primitive notion).

The primitive signs of the system P are the following:

I. Constants: "$\sim$" (not), "$\vee$" (or), "Π" (for all), "0" (zero), "f" (the successor of), "$($", "$)$" (parentheses);

II. Variables of type 1 (for individuals, that is, natural numbers including 0): "x_1", "y_1", "z_1", ...;

Variables of type 2 (for classes of individuals): "x_2", "y_2", "z_2", ...;

Variables of type 3 (for classes of classes of individuals): "x_3", "y_3", "z_3", ...;

And so on, for every natural number as a type.[17]

Remark: Variables for functions of two or more argument places (relations) need not be included among the primitive signs since we can define relations to be classes of ordered pairs, and ordered pairs to be classes of classes; for example, the ordered pair a, b can be defined to be $((a), (a, b))$, where (x, y) denotes the class whose sole elements are x and y, and (x) the class whose sole element is x.[18]

By a *sign of type* 1 we understand a combination of signs that has [any one of] the forms

$$a, fa, ffa, fffa, \ldots, \text{and so on,}$$

where a is either 0 or a variable of type 1. In the first case, we call such a sign a *numeral*. For $n > 1$ we understand by a *sign of type n* the same thing as by a *variable of type n*. A combination of signs that has the form $a(b)$, where b is a sign of type n

---

[16] The addition of the Peano axioms, as well as all other modifications introduced in the system PM, merely serves to simplify the proof and is dispensable in principle.

[17] It is assumed that we have denumerably many signs at our disposal for each type of variables.

[18] Nonhomogeneous relations, too, can be defined in this manner; for example, a relation between individuals and classes can be defined to be a class of elements of the form $((x_2), ((x_1), x_2))$. Every proposition about relations that is provable in PM is provable also when treated in this manner, as is readily seen.

and a a sign of type $n + 1$, will be called an *elementary formula*. We define the class of *formulas* to be the smallest class[19] containing all elementary formulas and containing $\sim(a)$, $(a) \vee (b)$, $x\Pi(a)$ (where x may be any variable)[18a] whenever it contains a and b. We call $(a) \vee (b)$ the *disjunction* of a and b, $\sim(a)$ the *negation* and $x\Pi(a)$ a *generalization* of a. A formula in which no free variable occurs (*free variable* being defined in the well-known manner) is called a *sentential formula* [[Satzformel]]. A formula with exactly n free individual variables (and no other free variables) will be called an *n-place relation sign*; for $n = 1$ it will also be called a *class sign*.

By Subst $a\binom{v}{b}$ (where a stands for a formula, v for a variable, and b for a sign of the same type as v) we understand the formula that results from a if in a we replace v, wherever it is free, by b.[20] We say that a formula a is a *type elevation* of another formula b if a results from b when the type of each variable occurring in b is increased by the same number.

The following formulas (I–V) are called *axioms* (we write them using these abbreviations, defined in the well-known manner: . , $\supset$, $\equiv$, (Ex), $=$,[21] and observing the usual conventions about omitting parentheses):[22]

I. 1. $\sim(fx_1 = 0)$,
   2. $fx_1 = fy_1 \supset x_1 = y_1$,
   3. $x_2(0) . x_1\Pi(x_2(x_1) \supset x_2(fx_1)) \supset x_1\Pi(x_2(x_1))$.

II. All formulas that result from the following schemata by substitution of any formulas whatsoever for p, q, r:

   1. $p \vee p \supset p$,   3. $p \vee q \supset q \vee p$,
   2. $p \supset p \vee q$,   4. $(p \supset q) \supset (r \vee p \supset r \vee q)$.

III. Any formula that results from either one of the two schemata

   1. $v\Pi(a) \supset$ Subst $a\binom{v}{c}$,
   2. $v\Pi(b \vee a) \supset b \vee v\Pi(a)$

when the following substitutions are made for a, v, b, and c (and the operation indicated by "Subst" is performed in 1):

For a any formula, for v any variable, for b any formula in which v does not occur free, and for c any sign of the same type as v, provided c does not contain any variable that is bound in a at a place where v is free.[23]

---

[19] Concerning this definition (and similar definitions occurring below) see *Łukasiewicz and Tarski 1930*.

[18a] Hence $x\Pi(a)$ is a formula even if x does not occur in a or is not free in a. In this case, of course, $x\Pi(a)$ means the same thing as a.

[20] In case v does not occur in a as a free variable we put Subst $a\binom{v}{b} = a$. Note that "Subst" is a metamathematical sign.

[21] $x_1 = y_1$ is to be regarded as defined by $x_2\Pi(x_2(x_1) \supset x_2(y_1))$, as in *PM* (I, *13) similarly for higher types).

[22] In order to obtain the axioms from the schemata listed we must therefore
(1) Eliminate the abbreviations and
(2) Add the omitted parentheses
(in II, III, and IV after carrying out the substitutions allowed).
Note that all expressions thus obtained are "formulas" in the sense specified above. (See also the exact definitions of the metamathematical notions on pp. 603–606.)

[23] Therefore c is a variable or 0 or a sign of the form $f \ldots fu$, where u is either 0 or a variable of type 1. Concerning the notion "free (bound) at a place in a", see I A 5 in *von Neumann 1927*.

IV. Every formula that results from the schema

1. $(Eu)(v\Pi(u(v) \equiv a))$

when for v we substitute any variable of type n, for u one of type $n + 1$, and for a any formula that does not contain u free. This axiom plays the role of the axiom of reducibility (the comprehension axiom of set theory).

V. Every formula that results from

1. $x_1\Pi(x_2(x_1) \equiv y_2(x_1)) \supset x_2 = y_2$

by type elevation (as well as this formula itself). This axiom states that a class is **completely determined by its elements**.

A formula c is called an *immediate consequence* of a and b if a is the formula $(\sim(b)) \vee (c)$, and it is called an *immediate consequence* of a if it is the formula $v\Pi(a)$, where v denotes any variable. The class of *provable formulas* is defined to be the smallest class of formulas that contains the axioms and is closed under the relation "immediate consequence".[24]

We now assign natural numbers to the primitive signs of the system P by the following one-to-one correspondence:

"0" ... 1   "$\sim$" ... 5   "Π" ... 9
"f" ... 3   "$\vee$" ... 7   "(" ... 11
                                     ")" ... 13;

to the variables of type n we assign the numbers of the form p^n (where p is a prime number > 13). Thus we have a one-to-one correspondence by which a finite sequence of natural numbers is associated with every finite sequence of primitive signs (hence also with every formula). We now map the finite sequences of natural numbers on natural numbers (again by a one-to-one correspondence), associating the number $2^{n_1} \cdot 3^{n_2} \cdot \ldots \cdot p_k^{n_k}$, where p_k denotes the kth prime number (in order of increasing magnitude), with the sequence $n_1, n_2, \ldots, n_k$. A natural number [out of a certain subset] is thus assigned one-to-one not only to every primitive sign but also to every finite sequence of such signs. We denote by $\Phi(a)$ the number assigned to the primitive sign (or to the sequence of primitive signs) a. Now let some relation (or class) $R(a_1, a_2, \ldots, a_n)$ between [or of] primitive signs or sequences of primitive signs be given. With it we associate the relation (or class) $R'(x_1, x_2, \ldots, x_n)$ between [or of] natural numbers that obtains between $x_1, x_2, \ldots, x_n$ if and only if there are some $a_1, a_2, \ldots, a_n$ such that $x_i = \Phi(a_i)$ $(i = 1, 2, \ldots, n)$ and $R(a_1, a_2, \ldots, a_n)$ hold. The relations between (or classes of) natural numbers that in this manner are associated with the metamathematical notions defined so far, for example, "variable", "formula", "sentential formula", "axiom", "provable formula", and so on, will be denoted by the same words in SMALL CAPITALS. The proposition that there are undecidable problems in the system P, for example, reads thus: There are SENTENTIAL FORMULAS a such that neither a nor the NEGATION of a is a PROVABLE FORMULA.

We now insert a parenthetic consideration that for the present has nothing to do

---

[24] The rule of substitution is rendered superfluous by the fact that all possible substitutions have already been carried out in the axioms themselves. (This procedure was used also in *von Neumann 1927*.)

with the formal system P. First we give the following definition: A number-theoretic function[25] $\varphi(x_1, x_2, \ldots, x_n)$ is said to be *recursively defined in terms of* the number-theoretic functions $\psi(x_1, x_2, \ldots, x_{n-1})$ and $\mu(x_1, x_2, \ldots, x_{n+1})$ if

$$\varphi(0, x_2, \ldots, x_n) = \psi(x_2, \ldots, x_n),$$
$$\varphi(k + 1, x_2, \ldots, x_n) = \mu(k, \varphi(k, x_2, \ldots, x_n), x_2, \ldots, x_n) \quad (2)$$

hold for all $x_2, \ldots, x_n, k$.[26]

A number-theoretic function φ is said to be *recursive* if there is a finite sequence of number-theoretic functions $\varphi_1, \varphi_2, \ldots, \varphi_n$ that ends with φ and has the property that every function φ_k of the sequence is recursively defined in terms of two of the preceding functions, or results from any of the preceding functions by substitution,[27] or, finally, is a constant or the successor function $x + 1$. The length of the shortest sequence of φ_i corresponding to a recursive function φ is called its *degree*. A relation $R(x_1, \ldots, x_n)$ between natural numbers is said to be *recursive*[28] if there is a recursive function $\varphi(x_1, \ldots, x_n)$ such that, for all $x_1, x_2, \ldots, x_n$,

$$R(x_1, \ldots, x_n) \sim [\varphi(x_1, \ldots, x_n) = 0].^{29}$$

The following theorems hold:

I. *Every function (relation) obtained from recursive functions (relations) by substitution of recursive functions for the variables is recursive; so is every function obtained from recursive functions by recursive definition according to schema (2);*

II. *If R and S are recursive relations, so are $\bar{R}$ and $R \lor S$ (hence also $R \, \& \, S$);*

III. *If the functions $\varphi(\mathfrak{x})$ and $\psi(\mathfrak{y})$ are recursive, so is the relation $\varphi(\mathfrak{x}) = \psi(\mathfrak{y})$;*[30]

IV. *If the function $\varphi(\mathfrak{x})$ and the relation $R(x, \mathfrak{y})$ are recursive, so are the relations S and T defined by*

$$S(\mathfrak{x}, \mathfrak{y}) \sim (Ex)[x \leq \varphi(\mathfrak{x}) \, \& \, R(x, \mathfrak{y})]$$

and

$$T(\mathfrak{x}, \mathfrak{y}) \sim (x)[x \leq \varphi(\mathfrak{x}) \to R(x, \mathfrak{y})],$$

as well as the function ψ defined by

$$\psi(\mathfrak{x}, \mathfrak{y}) = \varepsilon x[x \leq \varphi(\mathfrak{x}) \, \& \, R(x, \mathfrak{y})],$$

where $\varepsilon x F(x)$ means the least number x for which $F(x)$ holds and 0 in case there is no such number.

[25] That is, its domain of definition is the class of nonnegative integers (or of n-tuples of nonnegative integers) and its values are nonnegative integers.

[26] In what follows, lower-case italic letters (with or without subscripts) are always variables for nonnegative integers (unless the contrary is expressly noted).

[27] More precisely, by substitution of some of the preceding functions at the argument places of one of the preceding functions, for example, $\varphi_k(x_1, x_2) = \varphi_p[\varphi_q(x_1, x_2), \varphi_r(x_2)]$ ($p, q, r < k$). Not all variables on the left side need occur on the right side (the same applies to the recursion schema (2)).

[28] We include classes among relations (as one-place relations). Recursive relations R, of course, have the property that for every given n-tuple of numbers it can be decided whether $R(x_1, \ldots, x_n)$ holds or not.

[29] Whenever formulas are used to express a meaning (in particular, in all formulas expressing metamathematical propositions or notions), Hilbert's symbolism is employed. See *Hilbert and Ackermann 1928*.

[30] We use German letters, $\mathfrak{x}$, $\mathfrak{y}$, as abbreviations for arbitrary n-tuples of variables, for example, $x_1, x_2, \ldots, x_n$.

Theorem I follows at once from the definition of "recursive". Theorems II and III are consequences of the fact that the number-theoretic functions

$$\alpha(x), \quad \beta(x, y), \quad \gamma(x, y),$$

corresponding to the logical notions $\overline{}$, $\vee$, and $=$, namely,

$$\alpha(0) = 1, \alpha(x) = 0 \text{ for } x \neq 0,$$
$$\beta(0, x) = \beta(x, 0) = 0, \quad \beta(x, y) = 1 \text{ when } x \text{ and } y \text{ are both } \neq 0,$$
$$\gamma(x, y) = 0 \text{ when } x = y, \quad \gamma(x, y) = 1 \text{ when } x \neq y,$$

are recursive, as we can readily see. The proof of Theorem IV is briefly as follows. By assumption there is a recursive $\rho(x, \mathfrak{y})$ such that

$$R(x, \mathfrak{y}) \sim [\rho(x, \mathfrak{y}) = 0].$$

We now define a function $\chi(x, \mathfrak{y})$ by the recursion schema (2) in the following way:

$$\chi(0, \mathfrak{y}) = 0,$$
$$\chi(n + 1, \mathfrak{y}) = (n + 1).a + \chi(n, \mathfrak{y}).\alpha(a),{}^{31}$$

where $a = \alpha[\alpha(\rho(0, \mathfrak{y}))].\alpha[\rho(n + 1, \mathfrak{y})].\alpha[\chi(n, \mathfrak{y})]$. Therefore $\chi(n + 1, \mathfrak{y})$ is equal either to $n + 1$ (if $a = 1$) or to $\chi(n, \mathfrak{y})$ (if $a = 0$).[32] The first case clearly occurs if and only if all factors of a are 1, that is, if

$$\bar{R}(0, \mathfrak{y}) \ \& \ R(n + 1, \mathfrak{y}) \ \& \ [\chi(n, \mathfrak{y}) = 0]$$

holds. From this it follows that the function $\chi(n, \mathfrak{y})$ (considered as a function of n) remains 0 up to ⟦but not including⟧ the least value of n for which $R(n, \mathfrak{y})$ holds and, from there on, is equal to that value. (Hence, in case $R(0, \mathfrak{y})$ holds, $\chi(n, \mathfrak{y})$ is constant and equal to 0.) We have, therefore,

$$\psi(\mathfrak{x}, \mathfrak{y}) = \chi(\varphi(\mathfrak{x}), \mathfrak{y}),$$
$$S(\mathfrak{x}, \mathfrak{y}) \sim R[\psi(\mathfrak{x}, \mathfrak{y}), \mathfrak{y}].$$

The relation T can, by negation, be reduced to a case analogous to that of S. Theorem IV is thus proved.

The functions $x + y$, $x.y$, and x^y, as well as the relations $x < y$ and $x = y$, are recursive, as we can readily see. Starting from these notions, we now define a number of functions (relations) 1–45, each of which is defined in terms of preceding ones by the procedures given in Theorems I–IV. In most of these definitions several of the steps allowed by Theorems I–IV are condensed into one. Each of the functions (relations) 1–45, among which occur, for example, the notions "FORMULA", "AXIOM", and "IMMEDIATE CONSEQUENCE", is therefore recursive.

1. $x/y \equiv (Ez)[z \leq x \ \& \ x = y.z]$,[33]

x is divisible by y.[34]

---

[31] We assume familiarity with the fact that the functions $x + y$ (addition) and $x . y$ (multiplication) are recursive.

[32] a cannot take values other than 0 and 1, as can be seen from the definition of α.

[33] The sign $\equiv$ is used in the sense of "equality by definition"; hence in definitions it stands for either $=$ or $\sim$ (otherwise, the symbolism is Hilbert's).

[34] Wherever one of the signs (x), (Ex), or εx occurs in the definitions below, it is followed by a bound on x. This bound merely serves to ensure that the notion defined is recursive (see Theorem IV). But in most cases the *extension* of the notion defined would not change if this bound were omitted.

2. $\text{Prim}(x) \equiv \overline{(Ez)}[z \leq x \ \& \ z \neq 1 \ \& \ z \neq x \ \& \ x/z] \ \& \ x > 1$,

x is a prime number.

3. $0 \ Pr \ x \equiv 0$,

$(n + 1) \ Pr \ x \equiv \varepsilon y[y \leq x \ \& \ \text{Prim}(y) \ \& \ x/y \ \& \ y > n \ Pr \ x]$,

$n \ Pr \ x$ is the nth prime number (in order of increasing magnitude) contained in x.[34a]

4. $0! \equiv 1$,

$(n + 1)! \equiv (n + 1).n!$.

5. $Pr(0) \equiv 0$,

$Pr(n + 1) \equiv \varepsilon y[y \leq \{Pr(n)\}! + 1 \ \& \ \text{Prim}(y) \ \& \ y > Pr(n)]$,

$Pr(n)$ is the nth prime number (in order of increasing magnitude).

6. $n \ Gl \ x \equiv \varepsilon y[y \leq x \ \& \ x/(n \ Pr \ x)^y \ \& \ \overline{x/(n \ Pr \ x)^{y+1}}]$,

$n \ Gl \ x$ is the nth term of the number sequence assigned to the number x (for $n > 0$ and n not greater than the length of this sequence).

7. $l(x) \equiv \varepsilon y[y \leq x \ \& \ y \ Pr \ x > 0 \ \& \ (y + 1) \ Pr \ x = 0]$,

$l(x)$ is the length of the number sequence assigned to x.

8. $x*y \equiv \varepsilon z\{z \leq [Pr(l(x) + l(y))]^{x+y} \ \& \ (n)[n \leq l(x) \rightarrow n \ Gl \ z = n \ Gl \ x] \ \&$

$(n)[0 < n \leq l(y) \rightarrow (n + l(x)) \ Gl \ z = n \ Gl \ y]\}$,

$x*y$ corresponds to the operation of "concatenating" two finite number sequences.

9. $R(x) \equiv 2^x$,

$R(x)$ corresponds to the number sequence consisting of x alone (for $x > 0$).

10. $E(x) \equiv R(11)*x*R(13)$,

$E(x)$ corresponds to the operation of "enclosing within parentheses" (11 and 13 are assigned to the primitive signs "(" and ")", respectively).

11. $n \ \text{Var} \ x \equiv (Ez)[13 < z \leq x \ \& \ \text{Prim}(z) \ \& \ x = z^n] \ \& \ n \neq 0$,

x is a VARIABLE OF TYPE n.

12. $\text{Var}(x) \equiv (En)[n \leq x \ \& \ n \ \text{Var} \ x]$,

x is a VARIABLE.

13. $\text{Neg}(x) \equiv R(5)*E(x)$,

$\text{Neg}(x)$ is the NEGATION of x.

14. $x \ \text{Dis} \ y \equiv E(x)*R(7)*E(y)$,

$x \ \text{Dis} \ y$ is the DISJUNCTION of x and y.

15. $x \ \text{Gen} \ y \equiv R(x)*R(9)*E(y)$,

$x \ \text{Gen} \ y$ is the GENERALIZATION of y with respect to the VARIABLE x (provided x is a VARIABLE).

16. $0 \ N \ x \equiv x$,

$(n + 1) \ N \ x \equiv R(3)*n \ N \ x$,

$n \ N \ x$ corresponds to the operation of "putting the sign 'f' n times in front of x".

17. $Z(n) \equiv n \ N \ [R(1)]$,

$Z(n)$ is the NUMERAL denoting the number n.

18. $\text{Typ}'_1(x) \equiv (Em, n)\{m, n \leq x \ \& \ [m = 1 \ \lor \ 1 \ \text{Var} \ m] \ \& \ x = n \ N \ [R(m)]\}$,[34b]

x is a SIGN OF TYPE 1.

[34a] For $0 < n \leq z$, where z is the number of distinct prime factors of x. Note that $n \ Pr \ x = 0$ for $n = z + 1$.

[34b] $m, n \leq x$ stands for $m \leq x \ \& \ n \leq x$ (similarly for more than two variables).

19. $\text{Typ}_n(x) \equiv [n = 1 \,\&\, \text{Typ}'_1(x)] \lor [n > 1 \,\&\,$
    $(Ev)\{v \leq x \,\&\, n \text{ Var } v \,\&\, x = R(v)\}]$,

x is a SIGN OF TYPE n.

20. $\text{Elf}(x) \equiv (Ey, z, n)[y, z, n \leq x \,\&\, \text{Typ}_n(y) \,\&\,$
    $\text{Typ}_{n+1}(z) \,\&\, x = z*E(y)]$,

x is an ELEMENTARY FORMULA.

21. $Op(x, y, z) \equiv x = \text{Neg}(y) \lor x = y \text{ Dis } z \lor (Ev)[v \leq x \,\&\, \text{Var}(v) \,\&\,$
    $x = v \text{ Gen } y]$.

22. $FR(x) \equiv (n)\{0 < n \leq l(x) \rightarrow \text{Elf}(n \text{ Gl } x) \lor (Ep, q)[0 < p, q < n \,\&\,$
    $Op(n \text{ Gl } x, p \text{ Gl } x, q \text{ Gl } x)]\} \,\&\, l(x) > 0$,

x is a SEQUENCE OF FORMULAS, each of which either is an ELEMENTARY FORMULA or results from the preceding FORMULAS through the operations of NEGATION, DISJUNCTION, or GENERALIZATION.

23. $\text{Form}(x) \equiv (En)\{n \leq (Pr[l(x)^2])^{x \cdot [l(x)]^2} \,\&\, FR(n) \,\&\, x = [l(n)] \text{ Gl } n\}$,[35]

x is a FORMULA (that is, the last term of a FORMULA SEQUENCE n).

24. $v \text{ Geb } n, x \equiv \text{Var}(v) \,\&\, \text{Form}(x) \,\&\, (Ea, b, c)[a, b, c \leq x \,\&\,$
    $x = a*(v \text{ Gen } b)*c \,\&\, \text{Form}(b) \,\&\, l(a) + 1 \leq n \leq l(a) + l(v \text{ Gen } b)]$,

the VARIABLE v is BOUND in x at the nth place.

25. $v \text{ Fr } n, x \equiv \text{Var}(v) \,\&\, \text{Form}(x) \,\&\, v = n \text{ Gl } x \,\&\, n \leq l(x) \,\&\, \overline{v \text{ Geb } n, x}$,

the VARIABLE v is FREE in x at the nth place.

26. $v \text{ Fr } x \equiv (En)[n \leq l(x) \,\&\, v \text{ Fr } n, x]$,

v occurs as a FREE VARIABLE in x.

27. $Su \, x\binom{n}{y} \equiv \varepsilon z\{z \leq [Pr(l(x) + l(y))]^{x+y} \,\&\, [(Eu, v) \, u, v \leq x \,\&\,$
    $x = u*R(n \text{ Gl } x)*v \,\&\, z = u*y*v \,\&\, n = l(u) + 1]\}$,

$Su \, x\binom{n}{y}$ results from x when we substitute y for the nth term of x (provided that $0 < n \leq l(x)$).

28. $0 \, St \, v, x \equiv \varepsilon n\{n \leq l(x) \,\&\, v \text{ Fr } n, x \,\&\, \overline{(Ep)}[n < p \leq l(x) \,\&\, v \text{ Fr } p, x]\}$,
    $(k + 1) \, St \, v, x \equiv \varepsilon n\{n < k \, St \, v, x \,\&\, v \text{ Fr } n, x \,\&\, \overline{(Ep)}[n < p < k \, St \, v, x$
    $\,\&\, v \text{ Fr } p, x]\}$,

$k \, St \, v, x$ is the $(k + 1)$th place in x (counted from the right end of the FORMULA x) at which v is FREE in x (and 0 in case there is no such place).

29. $A(v, x) \equiv \varepsilon n\{n \leq l(x) \,\&\, n \, St \, v, x = 0\}$,

$A(v, x)$ is the number of places at which v is FREE in x.

30. $Sb_0(x_y^v) \equiv x$,
    $Sb_{k+1}(x_y^v) \equiv Su \, [Sb_k(x_y^v)](^{k \, St \, v, \, x}_{ y})$.

31. $Sb(x_y^v) \equiv Sb_{A(v,x)}(x_y^v)$,[36]

$Sb(x_y^v)$ is the notion SUBST $a\binom{v}{b}$ defined above.[37]

32. $x \text{ Imp } y \equiv [\text{Neg}(x)] \text{ Dis } y$,
    $x \text{ Con } y \equiv \text{Neg}\{[\text{Neg}(x)] \text{ Dis } [\text{Neg}(y)]\}$,

---

[35] That $n \leq (Pr([l(x)]^2))^{x \, [l(x)]^2}$ provides a bound can be seen thus: The length of the shortest sequence of formulas that corresponds to x can at most be equal to the number of subformulas of x. But there are at most $l(x)$ subformulas of length 1, at most $l(x) - 1$ of length 2, and so on, hence altogether at most $l(x)(l(x) + 1)/2 \leq [l(x)]^2$. Therefore all prime factors of n can be assumed to be less than $Pr([l(x)]^2)$, their number $\leq [(lx)]^2$, and their exponents (which are subformulas of x) $\leq x$.

[36] In case v is not a VARIABLE or x is not a FORMULA, $Sb(x_y^v) = x$.

[37] Instead of $Sb[Sb(x_y^v)_z^w]$ we write $Sb(x_y^v{}_z^w)$ (and similarly for more than two VARIABLES).

x Aeq $y \equiv (x \text{ Imp } y) \text{ Con } (y \text{ Imp } x)$,

v Ex $y \equiv \text{Neg}\{v \text{ Gen }[\text{Neg}(y)]\}$.

33. $n \text{ Th } x \equiv \varepsilon y \{y \leq x^{(x^n)} \& (k)[k \leq l(x) \to (k \text{ Gl } x \leq 13 \& k \text{ Gl } y = k \text{ Gl } x) \vee$
$(k \text{ Gl } x > 13 \& k \text{ Gl } y = k \text{ Gl } x.[1 \text{ Pr } (k \text{ Gl } x)]^n)]\}$,

$n \text{ Th } x$ is the nth TYPE ELEVATION of x (in case x and $n \text{ Th } x$ are FORMULAS).

Three specific numbers, which we denote by z_1, z_2, and z_3, correspond to the Axioms I, 1–3, and we define

34. $Z\text{-}Ax(x) \equiv (x = z_1 \vee x = z_2 \vee x = z_3)$.

35. $A_1\text{-}Ax(x) \equiv (Ey)[y \leq x \& \text{Form}(y) \& x = (y \text{ Dis } y) \text{ Imp } y]$,

x is a FORMULA resulting from Axiom schema II, 1 by substitution. Analogously, $A_2\text{-}Ax$, $A_3\text{-}Ax$, and $A_4\text{-}Ax$ are defined for Axioms [rather, Axiom Schemata] II, 2–4.

36. $A\text{-}Ax(x) \equiv A_1\text{-}Ax(x) \vee A_2\text{-}Ax(x) \vee A_3\text{-}Ax(x) \vee A_4\text{-}Ax(x)$,

x is a FORMULA resulting from a propositional axiom by substitution.

37. $Q(z, y, v) \equiv \overline{(En, m, w)}[n \leq l(y) \& m \leq l(z) \& w \leq z \&$
$w \equiv m \text{ Gl } z \& w \text{ Geb } n, y \& v \text{ Fr } n, y]$

z does not contain any VARIABLE BOUND in y at a place at which v is FREE.

38. $L_1\text{-}Ax(x) \equiv (Ev, y, z, n)\{v, y, z, n \leq x \& n \text{ Var } v \& \text{Typ}_n(z) \& \text{Form}(y) \&$
$Q(z, y, v) \& x = (v \text{ Gen } y) \text{ Imp } [Sb(y_z^v)]\}$,

x is a FORMULA resulting from Axiom schema III, 1 by substitution.

39. $L_2\text{-}Ax(x) \equiv (Ev, q, p)\{v, q, p \leq x \& \text{Var}(v) \& \text{Form}(p) \& \overline{v \text{ Fr } p} \& \text{Form}(q) \&$
$x = [v \text{ Gen } (p \text{ Dis } q)] \text{ Imp } [p \text{ Dis } (v \text{ Gen } q)]\}$,

x is a FORMULA resulting from Axiom schema III, 2 by substitution.

40. $R\text{-}Ax(x) \equiv (Eu, v, y, n)[u, v, y, n \leq x \& n \text{ Var } v \& (n + 1) \text{ Var } u \& \overline{u \text{ Fr } y} \&$
$\text{Form}(y) \& x = u \text{ Ex } \{v \text{ Gen } [[R(u) * E(R(v))] \text{ Aeq } y]\}]$,

x is a FORMULA resulting from Axiom schema IV, 1 by substitution.

A specific number z_4 corresponds to Axiom V, 1, and we define:

41. $M\text{-}Ax(x) \equiv (En)[n \leq x \& x = n \text{ Th } z_4]$.

42. $Ax(x) \equiv Z\text{-}Ax(x) \vee A\text{-}Ax(x) \vee L_1\text{-}Ax(x) \vee L_2\text{-}Ax(x) \vee R\text{-}Ax(x) \vee M\text{-}Ax(x)$,

x is an AXIOM.

43. $Fl(x, y, z) \equiv y = z \text{ Imp } x \vee (Ev)[v \leq x \& \text{Var}(v) \& x = v \text{ Gen } y]$,

x is an IMMEDIATE CONSEQUENCE of y and z.

44. $Bw(x) \equiv (n)\{0 < n \leq l(x) \to Ax(n \text{ Gl } x) \vee (Ep, q)[0 < p, q < n \&$
$Fl(n \text{ Gl } x, p \text{ Gl } x, q \text{ Gl } x)]\} \& l(x) > 0$,

x is a PROOF ARRAY (a finite sequence of FORMULAS, each of which is either an AXIOM or an IMMEDIATE CONSEQUENCE of two of the preceding FORMULAS.

45. $x B y \equiv Bw(x) \& [l(x)] \text{ Gl } x = y$,

x is a PROOF of the FORMULA y.

46. $Bew(x) \equiv (Ey) y B x$,

x is a PROVABLE FORMULA. ($Bew(x)$ is the only one of the notions 1–46 of which we cannot assert that it is recursive.)

The fact that can be formulated vaguely by saying: every recursive relation is definable in the system P (if the usual meaning is given to the formulas of this system), is expressed in precise language, *without* reference to any interpretation of the formulas of P, by the following theorem:

Theorem V. *For every recursive relation $R(x_1, \ldots, x_n)$ there exists an n-place*

RELATION SIGN r (*with the* FREE VARIABLES[38] $u_1, u_2, \ldots, u_n$) *such that for all n-tuples of numbers* $(x_1, \ldots, x_n)$ *we have*

$$R(x_1, \ldots, x_n) \to \text{Bew}[Sb(r_{Z(x_1)\ldots Z(x_n)}^{u_1 \ldots u_n})], \tag{3}$$

$$\bar{R}(x_1, \ldots, x_n) \to \text{Bew}[\text{Neg}(Sb(r_{Z(x_1)\ldots Z(x_n)}^{u_1 \ldots u_n}))]. \tag{4}$$

We shall give only an outline of the proof of this theorem because the proof does not present any difficulty in principle and is rather long.[39] We prove the theorem for all relations $R(x_1, \ldots, x_n)$ of the form $x_1 = \varphi(x_2, \ldots, x_n)$[40] (where φ is a recursive function) and we use induction on the degree of φ. For functions of degree 1 (that is, constants and the function $x + 1$) the theorem is trivial. Assume now that φ is of degree m. It results from functions of lower degrees, $\varphi_1, \ldots, \varphi_k$, through the operations of substitution or recursive definition. Since by the induction hypothesis everything has already been proved for $\varphi_1, \ldots, \varphi_k$, there are corresponding RELATION SIGNS, $r_1, \ldots, r_k$, such that (3) and (4) hold. The processes of definition by which φ results from $\varphi_1, \ldots, \varphi_k$ (substitution and recursive definition) can both be formally reproduced in the system P. If this is done, a new RELATION SIGN r is obtained from $r_1, \ldots, r_k$,[41] and, using the induction hypothesis, we can prove without difficulty that (3) and (4) hold for it. A RELATION SIGN r assigned to a recursive relation[42] by this procedure will be said to be recursive.

We now come to the goal of our discussions. Let κ be any class of FORMULAS. We denote by $\text{Flg}(\kappa)$ (the set of consequences of κ) the smallest set of FORMULAS that contains all FORMULAS of κ and all AXIOMS and is closed under the relation "IMMEDIATE CONSEQUENCE". κ is said to be ω-consistent if there is no CLASS SIGN a such that

$$(n)[Sb(a_{Z(n)}^v) \; \varepsilon \; \text{Flg}(\kappa)] \; \& \; [\text{Neg}(v \; \text{Gen} \; a)] \; \varepsilon \; \text{Flg}(\kappa),$$

where v is the FREE VARIABLE of the CLASS SIGN a.

Every ω-consistent system, of course, is consistent. As will be shown later, however, the converse does not hold.

The general result about the existence of undecidable propositions reads as follows:

Theorem VI. *For every ω-consistent recursive class κ of* FORMULAS *there are recursive* CLASS SIGNS r *such that neither v* Gen r *nor* Neg(v Gen r) *belongs to* Flg(κ) (*where v is the* FREE VARIABLE *of r*).

Proof. Let κ be any recursive ω-consistent class of FORMULAS. We define

$$Bw_\kappa(x) \equiv (n)[n \leq l(x) \to Ax(n \; Gl \; x) \lor (n \; Gl \; x) \; \varepsilon \; \kappa \; \lor$$

$$(Ep, q)\{0 < p, q < n \; \& \; Fl(n \; Gl \; x, p \; Gl \; x, q \; Gl \; x)\}] \; \& \; l(x) > 0 \tag{5}$$

---

[38] The VARIABLES $u_1, \ldots, u_n$ can be chosen arbitrarily. For example, there always is an r with the FREE VARIABLES 17, 19, 23, ..., and so on, for which (3) and (4) hold.

[39] Theorem V, of course, is a consequence of the fact that in the case of a recursive relation R it can, for every n-tuple of numbers, be decided *on the basis of the axioms of the system P* whether the relation R obtains or not.

[40] From this it follows at once that the theorem holds for every recursive relation, since any such relation is equivalent to $0 = \varphi(x_1, \ldots, x_n)$, where φ is recursive.

[41] When this proof is carried out in detail, r, of course, is not defined indirectly with the help of its meaning but in terms of its purely formal structure.

[42] Which, therefore, in the usual interpretation expresses the fact that this relation holds.

(see the analogous notion 44),

$$x \, B_\kappa \, y \equiv Bw_\kappa(x) \, \& \, [l(x)] \, Gl \, x = y \tag{6}$$
$$\text{Bew}_\kappa(x) \equiv (Ey) y \, B_\kappa \, x \tag{6.1}$$

(see the analogous notions 45 and 46).

We obviously have

$$(x)[\text{Bew}_\kappa(x) \sim x \, \varepsilon \, \text{Flg}(\kappa)] \tag{7}$$

and

$$(x)[\text{Bew}(x) \to \text{Bew}_\kappa(x)]. \tag{8}$$

We now define the relation

$$Q(x, y) \equiv \overline{x \, B_\kappa \, [Sb(y^{19}_{Z(y)})]}. \tag{8.1}$$

Since $x \, B_\kappa \, y$ (by (6) and (5)) and $Sb(y^{19}_{Z(y)})$ (by Definitions 17 and 31) are recursive, so is $Q(x, y)$. Therefore, by Theorem V and (8) there is a RELATION SIGN q (with the FREE VARIABLES 17 and 19) such that

$$\overline{x \, B_\kappa \, [Sb(y^{19}_{Z(y)})]} \to \text{Bew}_\kappa[Sb(q^{17 \, 19}_{Z(x) \, Z(y)})], \tag{9}$$

and

$$x \, B_\kappa \, [Sb(y^{19}_{Z(y)})] \to \text{Bew}_\kappa[Neg(Sb(q^{17 \, 19}_{Z(x) \, Z(y)}))]. \tag{10}$$

We put

$$p = 17 \, \text{Gen} \, q \tag{11}$$

(p is a CLASS SIGN with the FREE VARIABLE 19) and

$$r = Sb(q^{19}_{Z(p)}) \tag{12}$$

(r is a recursive CLASS SIGN[43] with the FREE VARIABLE 17).
Then we have

$$Sb(p^{19}_{Z(p)}) = Sb([17 \, \text{Gen} \, q]^{19}_{Z(p)}) = 17 \, \text{Gen} \, Sb(q^{19}_{Z(p)}) = 17 \, \text{Gen} \, r \tag{13}$$

(by (11) and (12));[44] furthermore

$$Sb(q^{17 \, 19}_{Z(x) \, Z(p)}) = Sb(r^{17}_{Z(x)}) \tag{14}$$

(by (12)). If we now substitute p for y in (9) and (10) and take (13) and (14) into account, we obtain

$$\overline{x \, B_\kappa \, (17 \, \text{Gen} \, r)} \to \text{Bew}_\kappa[Sb(r^{17}_{Z(x)})], \tag{15}$$

$$x \, B_\kappa \, (17 \, \text{Gen} \, r) \to \text{Bew}_\kappa[Neg(Sb(r^{17}_{Z(x)}))]. \tag{16}$$

This yields:

1. 17 Gen r is not κ-PROVABLE.[45] For, if it were, there would (by (6.1)) be an n such

---

[43] Since r is obtained from the recursive RELATION SIGN q through the replacement of a VARIABLE by a definite number, p. [Precisely stated the final part of this footnote (which refers to a side remark unnecessary for the proof) would read thus: "REPLACEMENT of a VARIABLE by the NUMERAL for p."]

[44] The operations Gen and Sb, of course, can always be interchanged in case they refer to different VARIABLES.

[45] By "x is κ-provable" we mean $x \, \varepsilon \, \text{Flg}(\kappa)$, which, by (7), means the same thing as $\text{Bew}_\kappa(x)$.

that $n\,B_\kappa$ (17 Gen r). Hence by (16) we would have $\text{Bew}_\kappa[\text{Neg}(Sb(r_{Z(n)}^{17}))]$, while, on the other hand, from the κ-PROVABILITY of 17 Gen r that of $Sb(r_{Z(n)}^{17})$ follows. Hence, κ would be inconsistent (and a fortiori ω-inconsistent).

2. Neg(17 Gen r) is not κ-PROVABLE. Proof: As has just been proved, 17 Gen r is not κ-PROVABLE; that is (by (6.1)), $(n)\overline{n\,B_\kappa\,(17\,\text{Gen}\,r)}$ holds. From this, $(n)\text{Bew}_\kappa[Sb(r_{Z(n)}^{17})]$ follows by (15), and that, in conjunction with $\text{Bew}_\kappa[\text{Neg}(17\,\text{Gen}\,r)]$, is incompatible with the ω-consistency of κ.

17 Gen r is therefore undecidable on the basis of κ, which proves Theorem VI.

We can readily see that the proof just given is constructive;[45a] that is, the following has been proved in an intuitionistically unobjectionable manner: Let an arbitrary recursively defined class κ of FORMULAS be given. Then, if a formal decision (on the basis of κ) of the SENTENTIAL FORMULA 17 Gen r (which [for each κ] can actually be exhibited) is presented to us, we can actually give

1. A PROOF of Neg(17 Gen r);
2. For any given n, a PROOF of $Sb(r_{Z(n)}^{17})$.

That is, a formal decision of 17 Gen r would have the consequence that we could actually exhibit an ω-inconsistency.

We shall say that a relation between (or a class of) natural numbers $R(x_1, \ldots, x_n)$ is *decidable* [[*entscheidungsdefinit*]] if there exists an n-place RELATION SIGN r such that (3) and (4) (see Theorem V) hold. In particular, therefore, by Theorem V every recursive relation is decidable. Similarly, a RELATION SIGN will be said to be *decidable* if it corresponds in this way to a decidable relation. Now it suffices for the existence of undecidable propositions that the class κ be ω-consistent and decidable. For the decidability carries over from κ to $x\,B_\kappa\,y$ (see (5) and (6)) and to $Q(x, y)$ (see (8.1)), and only this was used in the proof given above. In this case the undecidable proposition has the form v Gen r, where r is a decidable CLASS SIGN. (Note that it even suffices that κ be decidable in the system enlarged by κ.)

If, instead of assuming that κ is ω-consistent, we assume only that it is consistent, then, although the existence of an undecidable proposition does not follow [by the argument given above], it does follow that there exists a property (r) for which it is possible neither to give a counterexample nor to prove that it holds of all numbers. For in the proof that 17 Gen r is not κ-PROVABLE only the consistency of κ was used (see p. 608). Moreover from $\overline{\text{Bew}_\kappa}(17\,\text{Gen}\,r)$ it follows by (15) that, for every number x, $Sb(r_{Z(x)}^{17})$ is κ-PROVABLE and consequently that $\text{Neg}(Sb(r_{Z(x)}^{17}))$ is not κ-PROVABLE for any number.

If we adjoin Neg(17 Gen r) to κ, we obtain a class of FORMULAS κ′ that is consistent but not ω-consistent. κ′ is consistent, since otherwise 17 Gen r would be κ-PROVABLE. However, κ′ is not ω-consistent, because, by $\overline{\text{Bew}_\kappa}(17\,\text{Gen}\,r)$ and (15), $(x)\text{Bew}_\kappa Sb(r_{Z(x)}^{17})$ and, a fortiori, $(x)\text{Bew}_{\kappa'} Sb(r_{Z(x)}^{17})$ hold, while on the other hand, of course, $\text{Bew}_{\kappa'}[\text{Neg}(17\,\text{Gen}\,r)]$ holds.[46]

We have a special case of Theorem VI when the class κ consists of a finite number of FORMULAS (and, if we so desire, of those resulting from them by TYPE ELEVATION).

[45a] Since all existential statements occurring in the proof are based upon Theorem V, which, as is easily seen, is unobjectionable from the intuitionistic point of view.

[46] Of course, the existence of classes κ that are consistent but not ω-consistent is thus proved only on the assumption that there exists some consistent κ (that is, that P is consistent).

Every finite class κ is, of course, recursive.⁴⁶ᵃ Let a be the greatest number contained in κ. Then we have for κ

$$x \, \varepsilon \, \kappa \sim (Em, n)[m \leq x \,\&\, n \leq a \,\&\, n \, \varepsilon \, \kappa \,\&\, x = m \, Th \, n].$$

Hence κ is recursive. This allows us to conclude, for example, that, even with the help of the axiom of choice (for all types) or the generalized continuum hypothesis, not all propositions are decidable, provided these hypotheses are ω-consistent.

In the proof of Theorem VI no properties of the system P were used besides the following:

1. The class of axioms and the rules of inference (that is, the relation "immediate consequence") are recursively definable (as soon as we replace the primitive signs in some way by natural numbers);
2. Every recursive relation is definable (in the sense of Theorem V) in the system P.

Therefore, in every formal system that satisfies the assumptions 1 and 2 and is ω-consistent there are undecidable propositions of the form $(x)F(x)$, where F is a recursively defined property of natural numbers, and likewise in every extension of such a system by a recursively definable ω-consistent class of axioms. As can easily be verified, included among the systems satisfying the assumptions 1 and 2 are the Zermelo-Fraenkel and the von Neumann axiom systems of set theory,[47] as well as the axiom system of number theory consisting of the Peano axioms, recursive definition (by schema (2)), and the rules of logic.[48] Assumption 1 is satisfied by any system that has the usual rules of inference and whose axioms (like those of P) result from a finite number of schemata by substitution.⁴⁸ᵃ

3

We shall now deduce some consequences from Theorem VI, and to this end we give the following definition:

A relation (class) is said to be *arithmetic* if it can be defined in terms of the notions $+$ and $\cdot$ (addition and multiplication for natural numbers)[49] and the logical constants $\vee$, $\overline{}$, (x), and $=$, where (x) and $=$ apply to natural numbers only.[50] The notion "arithmetic proposition" is defined accordingly. The relations "greater than" and "congruent modulo n", for example, are arithmetic because we have

$$x > y \sim \overline{(Ez)}[y = x + z],$$
$$x \equiv y \pmod{n} \sim (Ez)[x = y + z.n \vee y = x + z.n].$$

---

⁴⁶ᵃ [[On page 190, lines 21, 22, and 23, of the German text the three occurrences of a are misprints and should be replaced by occurrences of κ.]]

[47] The proof of assumption 1 turns out to be even simpler here than for the system P, since there is just one kind of primitive variables (or two in von Neumann's system).

[48] See Problem III in *Hilbert 1928a*.

⁴⁸ᵃ As will be shown in Part II of this paper, the true reason for the incompleteness inherent in all formal systems of mathematics is that the formation of ever higher types can be continued into the transfinite (see *Hilbert 1925*, p. 184 [[above, p. 387]]), while in any formal system at most denumerably many of them are available. For it can be shown that the undecidable propositions constructed here become decidable whenever appropriate higher types are added (for example, the type ω to the system P). An analogous situation prevails for the axiom system of set theory.

[49] Here and in what follows, zero is always included among the natural numbers.

[50] The definiens of such a notion, therefore, must consist exclusively of the signs listed, variables for natural numbers, $x, y, \ldots$, and the signs 0 and 1 (variables for functions and sets are not permitted to occur). Instead of x any other number variable, of course, may occur in the prefixes.

We now have

Theorem VII. *Every recursive relation is arithmetic.*

We shall prove the following version of this theorem: every relation of the form $x_0 = \varphi(x_1, \ldots, x_n)$, where φ is recursive, is arithmetic, and we shall use induction on the degree of φ. Let φ be of degree s ($s > 1$). Then we have either

1. $\varphi(x_1, \ldots, x_n) = \rho[\chi_1(x_1, \ldots, x_n), \chi_2(x_1, \ldots, x_n), \ldots, \chi_m(x_1, \ldots, x_n)]$[51]

(where ρ and all χ_1 are of degrees less than s) or

2. $\varphi(0, x_2, \ldots, x_n) = \psi(x_2, \ldots, x_n)$,

$$\varphi(k + 1, x_2, \ldots, x_n) = \mu[k, \varphi(k, x_2, \ldots, x_n), x_2, \ldots, x_n]$$

(where ψ and μ are of degrees less than s).

In the first case we have

$$x_0 = \varphi(x_1, \ldots, x_n) \sim (Ey_1, \ldots, y_m)[R(x_0, y_1, \ldots, y_m) \ \& $$
$$S_1(y_1, x_1, \ldots, x_n) \ \& \ \ldots \ \& \ S_m(y_m, x_1, \ldots, x_n)],$$

where R and S_i are the arithmetic relations, existing by the induction hypothesis, that are equivalent to $x_0 = \rho(y_1, \ldots, y_m)$ and $y = \chi_i(x_1, \ldots, x_n)$, respectively. Hence in this case $x_0 = \varphi(x_1, \ldots, x_n)$ is arithmetic.

In the second case we use the following method. We can express the relation $x_0 = \varphi(x_1, \ldots, x_n)$ with the help of the notion "sequence of numbers" (f)[52] in the following way:

$$x_0 = \varphi(x_1, \ldots, x_n) \sim (Ef)\{f_0 = \psi(x_2, \ldots, x_n) \ \& \ (k)[k < x_1 \rightarrow$$
$$f_{k+1} = \mu(k, f_k, x_2, \ldots, x_n)] \ \& \ x_0 = f_{x_1}\}.$$

If $S(y, x_2, \ldots, x_n)$ and $T(z, x_1, \ldots, x_{n+1})$ are the arithmetic relations, existing by the induction hypothesis, that are equivalent to $y = \psi(x_2, \ldots, x_n)$ and $z = \mu(x_1, \ldots, x_{n+1})$, respectively, then

$$x_0 = \varphi(x_1, \ldots, x_n) \sim (Ef)\{S(f_0, x_2, \ldots, x_n) \ \& \ (k)[k < x_1 \rightarrow$$
$$T(f_{k+1}, k, f_k, x_2, \ldots, x_n)] \ \& \ x_0 = f_{x_1}\}. \qquad (17)$$

We now replace the notion "sequence of numbers" by "pair of numbers", assigning to the number pair n, d the number sequence $f^{(n,d)}$ ($f_k^{(n,d)} = [n]_{1+(k+1)d}$), where $[n]_p$ denotes the least nonnegative remainder of n modulo p.

We then have

Lemma 1. If f is any sequence of natural numbers and k any natural number, there exists a pair of natural numbers, n, d such that $f^{(n,d)}$ and f agree in the first k terms.

Proof. Let l be the maximum of the numbers $k, f_0, f_1, \ldots, f_{k-1}$. Let us determine an n such that

$$n \equiv f_i [\mathrm{mod}(1 + (i + 1)l!)] \quad \text{for } i = 0, 1, \ldots, k - 1,$$

which is possible, since any two of the numbers $1 + (i + 1)l!$ ($i = 0, 1, \ldots, k - 1$)

---

[51] Of course, not all $x_1, \ldots, x_n$ need occur in the χ_1 (see the example in footnote 27).

[52] f here is a variable with the [infinite] sequences of natural numbers as its domain of values. f_k denotes the $(k + 1)$th term of a sequence f (f_0 denoting the first).

are relatively prime. For a prime number contained in two of these numbers would also be contained in the difference $(i_1 - i_2)l!$ and therefore, since $|i_1 - i_2| < l$, in $l!$; but this is impossible. The number pair $n, l!$ then has the desired property.

Since the relation $x = [n]_p$ is defined by

$$x \equiv n \ (\mathrm{mod}\ p) \ \& \ x < p$$

and is therefore arithmetic, the relation $P(x_0, x_1, \ldots, x_n)$, defined as follows:

$$P(x_0, \ldots, x_n) \equiv (En, d)\{S([n]_{d+1}, x_2, \ldots, x_n) \ \& \ (k) \ [k < x_1 \rightarrow$$
$$T([n]_{1+d(k+2)}, k, [n]_{1+d(k+1)}, x_2, \ldots, x_n)] \ \& \ x_0 = [n]_{1+d(x_1+1)}\},$$

is also arithmetic. But by (17) and Lemma 1 it is equivalent to $x_0 = \varphi(x_1, \ldots, x_n)$ (the sequence f enters in (17) only through its first $x_1 + 1$ terms). Theorem VII is thus proved.

By Theorem VII, for every problem of the form $(x)F(x)$ (with recursive F) there is an equivalent arithmetic problem. Moreover, since the entire proof of Theorem VII (for every particular F) can be formalized in the system P, this equivalence is provable in P. Hence we have

Theorem VIII. *In any of the formal systems mentioned in Theorem VI[53] there are undecidable arithmetic propositions.*

By the remark on page 610, the same holds for the axiom system of set theory and its extensions by ω-consistent recursive classes of axioms.

Finally, we derive the following result:

Theorem IX. *In any of the formal systems mentioned in Theorem VI[53] there are undecidable problems of the restricted functional calculus[54]* (that is, formulas of the restricted functional calculus for which neither validity nor the existence of a counterexample is provable).[55]

This is a consequence of

Theorem X. *Every problem of the form $(x)F(x)$ (with recursive F) can be reduced to the question whether a certain formula of the restricted functional calculus is satisfiable* (that is, for every recursive F we can find a formula of the restricted functional calculus that is satisfiable if and only if $(x)F(x)$ is true.

By formulas of the restricted functional calculus (r. f. c.) we understand expressions formed from the primitive signs $\overline{}$, $\vee$, (x), $=$, $x, y, \ldots$ (individual variables), $F(x), G(x, y), H(x, y, z), \ldots$ (predicate and relation variables), where (x) and $=$ apply to individuals only.[56] To these signs we add a third kind of variables, $\varphi(x)$, $\psi(x, y)$,

---

[53] These are the ω-consistent systems that result from P when recursively definable classes of axioms are added.

[54] See *Hilbert and Ackermann 1928*.

In the system P we must understand by formulas of the restricted functional calculus those that result from the formulas of the restricted functional calculus of PM when relations are replaced by classes of higher types as indicated on page 599.

[55] In *1930a* I showed that every formula of the restricted functional calculus either can be proved to be valid or has a counterexample. However, by Theorem IX the existence of this counterexample is *not* always provable (in the formal systems we have been considering).

[56] Hilbert and Ackermann (*1928*) do not include the sign $=$ in the restricted functional calculus. But for every formula in which the sign $=$ occurs there exists a formula that does not contain this sign and is satisfiable if and only if the original formula is (see *Gödel 1930a*).

$\kappa(x, y, z)$, and so on, which stand for object-functions⟦Gegenstandsfunktionen⟧ (that is, $\varphi(x)$, $\psi(x, y)$, and so on denote single-valued functions whose arguments and values are individuals).[57] A formula that contains variables of the third kind in addition to the signs of the r. f. c. first mentioned will be called a formula in the extended sense (i. e. s.).[58] The notions "satisfiable" and "valid" carry over immediately to formulas i. e. s., and we have the theorem that, for any formula A i. e. s., we can find a formula B of the r. f. c. proper such that A is satisfiable if and only if B is. We obtain B from A by replacing the variables of the third kind, $\varphi(x)$, $\psi(x, y)$, ..., that occur in A with expressions of the form $(\imath z)F(z, x)$, $(\imath z)G(z, x, y)$, ..., by eliminating the "descriptive" functions by the method used in PM (I, *14), and by logically multiplying[59] the formula thus obtained by an expression stating about each $F, G, \ldots$ put in place of some $\varphi, \psi, \ldots$ that it holds for a unique value of the first argument [for any choice of values for the other arguments].

We now show that, for every problem of the form $(x)F(x)$ (with recursive F), there is an equivalent problem concerning the satisfiability of a formula i. e. s., so that, on account of the remark just made, Theorem X follows.

Since F is recursive, there is a recursive function $\Phi(x)$ such that $F(x) \sim [\Phi(x) = 0]$, and for Φ there is sequence of functions, $\Phi_1, \Phi_2, \ldots, \Phi_n$, such that $\Phi_n = \Phi$, $\Phi_1(x) = x + 1$, and for every Φ_k ($1 < k \leq n$) we have either

1. $$(x_2, \ldots, x_m)[\Phi_k(0, x_2, \ldots, x_m) = \Phi_p(x_2, \ldots, x_m)],$$
$$(x, x_2, \ldots, x_m)\{\Phi_k[\Phi_1(x), x_2, \ldots, x_m] = \Phi_q[x, \Phi_k(x, x_2, \ldots, x_m), x_2, \ldots, x_m]\}, \quad (18)$$

with $p, q < k$,[59a]

or

2. $$(x_1, \ldots, x_m)[\Phi_k(x_1, \ldots, x_m) = \Phi_r(\Phi_{i_1}(\mathfrak{x}_1), \ldots, \Phi_{i_s}(\mathfrak{x}_s))],^{60} \quad (19)$$

with $r < k$, $i_v < k$ (for $v = 1, 2, \ldots, s$),

or

3. $$(x_1, \ldots, x_m)[\Phi_k(x_1, \ldots, x_m) = \Phi_1(\Phi_1, \ldots, \Phi_1(0))]. \quad (20)$$

We then form the propositions

$$(x)\overline{\Phi_1(x) = 0} \ \& \ (x, y)[\Phi_1(x) = \Phi_1(y) \to x = y], \quad (21)$$

$$(x)[\Phi_n(x) = 0]. \quad (22)$$

In all of the formulas (18), (19), (20) (for $k = 2, 3, \ldots, n$) and in (21) and (22) we now replace the functions Φ_i by function variables φ_i and the number 0 by an

---

[57] Moreover, the domain of definition is always supposed to be the *entire* domain of individuals.

[58] Variables of the third kind may occur at all argument places occupied by individual variables, for example, $y = \varphi(x)$, $F(x, \varphi(y))$, $G(\psi(x, \varphi(y)), x)$, and the like.

[59] That is, by forming the conjunction.

[59a] [The last clause of footnote 27 was not taken into account in the formulas (18). But an explicit formulation of the cases with fewer variables on the right side is actually necessary here for the formal correctness of the proof, unless the identity function, $I(x) = x$, is added to the initial functions.]

[60] The $\mathfrak{x}_i$ ($i = 1, \ldots, s$) stand for finite sequences of the variables $x_1, x_2, \ldots, x_m$; for example, x_1, x_3, x_2.

individual variable x_0 not used so far, and we form the conjunction C of all the formulas thus obtained.

The formula $(Ex_0)C$ then has the required property, that is,

1. If $(x)[\Phi(x) = 0]$ holds, $(Ex_0)C$ is satisfiable. For the functions $\Phi_1, \Phi_2, \ldots, \Phi_n$ obviously yield a true proposition when substituted for $\varphi_1, \varphi_2, \ldots, \varphi_n$ in $(Ex_0)C$;

2. If $(Ex_0)C$ is satisfiable, $(x)[\Phi(x) = 0]$ holds.

Proof. Let $\psi_1, \psi_2, \ldots, \psi_n$ be the functions (which exist by assumption) that yield a true proposition when substituted for $\varphi_1, \varphi_2, \ldots, \varphi_n$ in $(Ex_0)C$. Let $\Im$ be their domain of individuals. Since $(Ex_0)C$ holds for the functions ψ_i, there is an individual a (in $\Im$) such that all of the formulas (18)–(22) go over into true propositions, (18′)–(22′), when the Φ_i are replaced by the ψ_i and 0 by a. We now form the smallest subclass of $\Im$ that contains a and is closed under the operation $\psi_1(x)$. This subclass ($\Im'$) has the property that every function ψ_i, when applied to elements of $\Im'$, again yields elements of $\Im'$. For this holds of ψ_1 by the definition of $\Im'$, and by (18′), (19′), and (20′) it carries over from ψ_i with smaller subscripts to ψ_i with larger ones. The functions that result from the ψ_i when these are restricted to the domain $\Im'$ of individuals will be denoted by ψ_i'. All of the formulas (18)–(22) hold for these functions also (when we replace 0 by a and Φ_i by ψ_i').

Because (21) holds for ψ_1' and a, we can map the individuals of $\Im'$ one-to-one onto the natural numbers in such a manner that a goes over into 0 and the function ψ_1' into the successor function Φ_1. But by this mapping the functions ψ_i' go over into the functions Φ_i, and, since (22) holds for ψ_n' and a, $(x)[\Phi_n(x) = 0]$, that is, $(x)[\Phi(x) = 0]$, holds, which was to be proved.[61]

Since (for each particular F) the argument leading to Theorem X can be carried out in the system P, it follows that any proposition of the form $(x)F(x)$ (with recursive F) can in P be proved equivalent to the proposition that states about the corresponding formula of the r. f. c. that it is satisfiable. Hence the undecidability of one implies that of the other, which proves Theorem IX.[62]

4

The results of Section 2 have a surprising consequence concerning a consistency proof for the system P (and its extensions), which can be stated as follows:

Theorem XI. *Let κ be any recursive consistent[63] class of* FORMULAS; *then the* SENTENTIAL FORMULA *stating that κ is consistent is not κ-*PROVABLE; *in particular, the consistency of P is not provable in P,[64] provided P is consistent (in the opposite case, of course, every proposition is provable [in P]).*

The proof (briefly outlined) is as follows. Let κ be some recursive class of FORMULAS chosen once and for all for the following discussion (in the simplest case it is the

---

[61] Theorem X implies, for example, that Fermat's problem and Goldbach's problem could be solved if the decision problem for the r. f. c. were solved.

[62] Theorem IX, of course, also holds for the axiom system of set theory and for its extensions by recursively definable ω-consistent classes of axioms, since there are undecidable propositions of the form $(x)F(x)$ (with recursive F) in these systems too.

[63] "κ is consistent" (abbreviated by "Wid(κ)") is defined thus: Wid(κ) $\equiv$ (Ex)(Form(x) & $\overline{\text{Bew}_\kappa(x)}$).

[64] This follows if we substitute the empty class of FORMULAS for κ.

empty class). As appears from 1, page 608, only the consistency of κ was used in proving that 17 Gen r is not κ-PROVABLE;[65] that is, we have

$$\mathrm{Wid}(\kappa) \to \overline{\mathrm{Bew}_\kappa}(17 \text{ Gen } r), \tag{23}$$

that is, by (6.1),

$$\mathrm{Wid}(\kappa) \to (x)\,\overline{x\,B_\kappa\,(17 \text{ Gen } r)}.$$

By (13), we have

$$17 \text{ Gen } r = Sb(p^{19}_{Z(p)}),$$

hence

$$\mathrm{Wid}(\kappa) \to (x)\,\overline{x\,B_\kappa\,Sb(p^{19}_{Z(p)})},$$

that is, by (8.1),

$$\mathrm{Wid}(\kappa) \to (x)Q(x, p). \tag{24}$$

We now observe the following: all notions defined (or statements proved) in Section 2,[66] and in Section 4 up to this point, are also expressible (or provable) in P. For throughout we have used only the methods of definition and proof that are customary in classical mathematics, as they are formalized in the system P. In particular, κ (like every recursive class) is definable in P. Let w be the SENTENTIAL FORMULA by which $\mathrm{Wid}(\kappa)$ is expressed in P. According to (8.1), (9), and (10), the relation $Q(x, y)$ is expressed by the RELATION SIGN q, hence $Q(x, p)$ by r (since, by (12), $r = Sb(q^{19}_{Z(p)})$), and the proposition $(x)Q(x, p)$ by 17 Gen r.

Therefore, by (24), w Imp (17 Gen r) is provable in P[67] (and a fortiori κ-PROVABLE). If now w were κ-PROVABLE, then 17 Gen r would also be κ-PROVABLE, and from this it would follow, by (23), that κ is not consistent.

Let us observe that this proof, too, is constructive; that is, it allows us to actually derive a contradiction from κ, once a PROOF of w from κ is given. The entire proof of Theorem XI carries over word for word to the axiom system of set theory, M, and to that of classical mathematics,[68] A, and here, too, it yields the result: There is no consistency proof for M, or for A, that could be formalized in M, or A, respectively, provided M, or A, is consistent. I wish to note expressly that Theorem XI (and the corresponding results for M and A) do not contradict Hilbert's formalistic viewpoint. For this viewpoint presupposes only the existence of a consistency proof in which nothing but finitary means of proof is used, and it is conceivable that there exist finitary proofs that *cannot* be expressed in the formalism of P (or of M or A).

Since, for any consistent class κ, w is not κ-PROVABLE, there always are propositions (namely w) that are undecidable (on the basis of κ) as soon as $\mathrm{Neg}(w)$ is not κ-PROVABLE; in other words, we can, in Theorem VI, replace the assumption of ω-consistency by the following: The proposition "κ is inconsistent" is not κ-PROVABLE. (Note that there are consistent κ for which this proposition is κ-PROVABLE.)

---

[65] Of course, r (like p) depends on κ.

[66] From the definition of "recursive" on page 602 to the proof of Theorem VI inclusive.

[67] That the truth of w Imp (17 Gen r) can be inferred from (23) is simply due to the fact that the undecidable proposition 17 Gen r asserts its own unprovability, as was noted at the very beginning.

[68] See *von Neumann 1927*.

In the present paper we have on the whole restricted ourselves to the system P, and we have only indicated the applications to other systems. The results will be stated and proved in full generality in a sequel to be published soon.[68a] In that paper, also, the proof of Theorem XI, only sketched here, will be given in detail.

Note added 28 August 1963. In consequence of later advances, in particular of the fact that due to A. M. Turing's work[69] a precise and unquestionably adequate definition of the general notion of formal system[70] can now be given, a completely general version of Theorems VI and XI is now possible. That is, it can be proved rigorously that in *every* consistent formal system that contains a certain amount of finitary number theory there exist undecidable arithmetic propositions and that, moreover, the consistency of any such system cannot be proved in the system.

[68a] [[This explains the "I" in the title of the paper. The author's intention was to publish this sequel in the next volume of the *Monatshefte*. The prompt acceptance of his results was one of the reasons that made him change his plan.]]

[69] See *Turing 1937*, p. 249.

[70] In my opinion the term "formal system" or "formalism" should never be used for anything but this notion. In a lecture at Princeton (mentioned in *Princeton University 1946*, p. 11 [[see *Davis 1965*, pp. 84–88]]) I suggested certain transfinite generalizations of formalisms, but these are something radically different from formal systems in the proper sense of the term, whose characteristic property is that reasoning in them, in principle, can be completely replaced by mechanical devices.

ON COMPLETENESS AND CONSISTENCY
(1931a)

Let Z be the formal system that we obtain by supplementing the Peano axioms with the schema of definition by recursion (on one variable) and the logical rules of the *restricted* functional calculus. Hence Z is to contain no variables other than variables for individuals (that is, natural numbers), and the principle of mathematical induction must therefore be formulated as a rule of inference. Then the following hold:

1. Given any formal system S in which there are finitely many axioms and in which the sole principles of inference are the rule of substitution and the rule of implication, if S contains[1] Z, S is incomplete, that is, there are in S propositions (in

---

[1] That a formal system S contains another formal system T means that every proposition expressible (provable) in T is expressible (provable) also in S.

[[Remark by the author, 18 May 1966:]]

[This definition is not precise, and, if made precise in the straightforward manner, it does not yield a sufficient condition for the nondemonstrability in S of the consistency of S. A sufficient condition is obtained if one uses the following definition: "S contains T if and only if every meaningful formula (or axiom or rule (of inference, of definition, or of construction of axioms)) of T *is a* meaningful formula (or axiom, and so forth) of S, that is, if S is an extension of T".

Under the weaker hypothesis that Z is recursively one-to-one translatable into S, with demonstrability preserved in this direction, the consistency, even of very strong systems S, *may* be provable in S and even in primitive recursive number theory. However, what can be shown to be unprovable in S is the fact that the rules of the equational calculus applied to equations, between primitive recursive terms, demonstrable in S yield only correct numerical equations (provided that S possesses the property that is asserted to be unprovable). Note that it is necessary to prove this "outer" consistency of S (which for the usual systems is trivially equivalent with consistency) in order to "justify", in the sense of Hilbert's program, the transfinite axioms of a

particular, propositions of Z) that are undecidable on the basis of the axioms of S, provided that S is ω-consistent. Here a system is said to be ω-consistent if, for no property F of natural numbers, $(Ex)\overline{Fx}$ as well as all the formulas $F(i)$, $i = 1, 2, \ldots$, are provable.

2. In particular, in every system S of the kind just mentioned the proposition that S is consistent (more precisely, the equivalent arithmetic proposition that we obtain by mapping the formulas one-to-one on natural numbers) is unprovable.

Theorems 1 and 2 hold also for systems in which there are infinitely many axioms and in which there are other principles of inference than those mentioned above, provided that when we enumerate the formulas (in order of increasing length and, for equal length, in lexicographical order) the class of numbers assigned to the axioms is definable and decidable [[entscheidungsdefinit]] in the system Z, and that the same holds of the following relation $R(x_1, x_2, \ldots, x_n)$ between natural numbers: "the formula with number x_1 follows from the formulas with numbers $x_2, \ldots, x_n$ by a single application of one of the rules of inference". Here a relation (class) $R(x_1, x_2, \ldots, x_n)$ is said to be decidable in Z if for every n-tuple $(k_1, k_2, \ldots, k_n)$ of natural numbers either $R(k_1, k_2, \ldots, k_n)$ or $\overline{R}(k_1, k_2, \ldots, k_n)$ is provable in Z. (At present no decidable number-theoretic relation is known that is not definable and decidable already in Z.)

If we imagine that the system Z is successively enlarged by the introduction of variables for classes of numbers, classes of classes of numbers, and so forth, together with the corresponding comprehension axioms, we obtain a sequence (continuable into the transfinite) of formal systems that satisfy the assumptions mentioned above, and it turns out that the consistency (ω-consistency) of any of those systems is provable in all subsequent systems. Also, the undecidable propositions constructed for the proof of Theorem 1 become decidable by the adjunction of higher types and the corresponding axioms; however, in the higher systems we can construct other undecidable propositions by the same procedure, and so forth. To be sure, all the propositions thus constructed are expressible in Z (hence are number-theoretic propositions); they are, however, not decidable in Z, but only in higher systems, for example, in that of analysis. In case we adopt a type-free construction of mathematics, as is done in the axiom system of set theory, axioms of cardinality (that is, axioms postulating the existence of sets of ever higher cardinality) take the place of the type extensions, and it follows that certain arithmetic propositions that are undecidable in Z become decidable by axioms of cardinality, for example, by the axiom that there exist sets whose cardinality is greater than every α_n, where $\alpha_0 = \aleph_0$, $\alpha_{n+1} = 2^{\alpha_n}$.

---

system S. ("Rules of the equational calculus" in the foregoing means the two rules of substituting primitive recursive terms for variables and substituting one such term for another to which it has been proved equal.)

The last-mentioned theorem and Theorem 1 of the paper remain valid for much weaker systems than Z, in particular for primitive recursive number theory, that is, what remains of Z if quantifiers are omitted. With insignificant changes in the wording of the conclusions of the two theorems they even hold for any recursive translation into S of the equations between primitive recursive terms, under the sole hypothesis of ω-consistency (or outer consistency) of S in this translation.]

REFERENCES
INDEX

References

Throughout this volume, an author's name followed by a year number, both in italics, denotes an entry in the present list of references. Thus "*Frege 1873*" denotes the first publication listed under Frege in this list. When the context leaves no doubt as to the author, the year number alone is used. For papers that originally were addresses or lectures, the year indicated is that in which they were delivered; for communications to learned societies, it is that in which they were made; for other papers, it is that of the complete volume of the periodical in which they appear (there are a few exceptions, with irregular publications—any compiler of a bibliography knows that no rule is without exception). For one author, additional titles in the same year are distinguished by *a*, *b*, and so on; an attempt has been made to follow the actual chronological order.

Ackermann, Wilhelm
- *1924* Begründung des "tertium non datur" mittels der Hilbertschen Theorie der Widerspruchsfreiheit, *Mathematische Annalen 93*, 1–36.
- *1928* Zum Hilbertschen Aufbau der reellen Zahlen, *ibid. 99*, 118–133; English translation in *van Heijenoort 1967*, 493–507.
- See Hilbert, David, and Wilhelm Ackermann.

Angelelli, Ignacio, and Terrell Ward Bynum
- *1966* Note on Frege's Begriffsschrift, *Notre Dame journal of formal logic 7*, 369–370.

Bauer-Mengelberg, Stefan
- *1965* Review of *Gödel 1962*, *The journal of symbolic logic 30*, 359–362.
- *1966* Review of Elliott Mendelson's translation of *Gödel 1931* in *Davis 1965*, *ibid. 31*, 486–489.

Benacerraf, Paul, and Hilary Putnam (eds.)
- *1964* Philosophy of mathematics: selected readings (Prentice-Hall, Englewood Cliffs, New Jersey).

Bernays, Paul
- *1923* Erwiderung auf die Note von Herrn Aloys Müller: "Über Zahlen als Zeichen", *Mathematische Annalen 90*, 159–163; reprinted in *Annalen der Philosophie und philosophischen Kritik 4* (1924), 492–497.
- *1938* Sur les questions méthodologiques actuelles de la théorie hilbertienne de la démonstration, in *Gonseth 1938*, 144–152; Discussion, 153–161.
- *1954* Zur Beurteilung der Situation in der beweistheoretischen Forschung, *Revue internationale de philosophie 8*, 9–13; Discussion, 15–21.
- See Hilbert, David, and Paul Bernays.

Bernoulli, Jakob
- *1686* Demonstratio rationum etc., *Acta eruditorum*, 360–361; reprinted in *Bernoulli 1744*, 282–283.
- *1744* Opera, vol. 1 (Geneva).

REFERENCES

Bynum, Terrell Ward
  See Angelelli, Ignacio, and Terrell Ward Bynum.

Church, Alonzo
  1936 A note on the Entscheidungsproblem, *The journal of symbolic logic 1*, 40–41; correction, *ibid.*, 101–102; reprinted in *Davis 1965*, 110–115.

Davis, Martin (ed.)
  1965 *The undecidable. Basic papers on undecidable propositions, unsolvable problems, and computable functions* (Raven Press, Hewlett, New York).

Dedekind, Richard
  1888 *Was sind und was sollen die Zahlen?* (Brunswick).
  1890 Über den Begriff des Unendlichen, unpublished manuscript in the Niedersächsische Staats- und Universitätsbibliothek, Göttingen (Cod. Ms. Dedekind 3, folder 1), dated 8 February 1890.
  1890a Letter to Keferstein, in *van Heijenoort 1967*, 98–103.
  1893 2nd. ed. of *Dedekind 1888*, unchanged but for an additional preface.
  1911 3rd. ed. of *Dedekind 1888*, unchanged but for a third preface.

Fraenkel, Abraham A.
  1927 *Zehn Vorlesungen über die Grundlegung der Mengenlehre* (Teubner, Leipzig and Berlin).

Frege, Gottlob
  1873 *Ueber eine geometrische Darstellung der imaginären Gebilde in der Ebene*, Inaugural-Dissertation der philosophischen Facultät zu Göttingen zur Erlangung der Doctorwürde (Jena).
  1874 *Rechnungsmethoden, die sich auf eine Erweiterung des Grössenbegriffs gründen*, Dissertation zur Erlangung der venia docendi bei der philosophischen Fakultät in Jena (Jena).
  1879 *Begriffsschrift, eine der arithmetischen nachgebildete Formelsprache des reinen Denkens* (Halle); reprinted in *Frege 1964*; English translation in *van Heijenoort 1967*, 1–82; 1–82 of the present volume.
  1879a Anwendungen der Begriffsschrift, *Sitzungsberichte der Jenaischen Gesellschaft für Medicin und Naturwissenschaft für das Jahr 1879*, 29–33; reprinted in *Frege 1964*, 89–93.
  1882 Ueber den Zweck der Begriffsschrift, *Sitzungsberichte der Jenaischen Gesellschaft für Medicin und Naturwissenschaft für das Jahr 1882* (1883), 1–10; reprinted in *Frege 1964*, 97–106.
  1882a Ueber die wissenschaftliche Berechtigung einer Begriffsschrift, *Zeitschrift für Philosophie und philosophische Kritik*, new series, *81*, 48–56; reprinted in *Frege 1962*, 89–95, and in *Frege 1964*, 106–114.
  1884 *Die Grundlagen der Arithmetik, eine logisch-mathematische Untersuchung über den Begriff der Zahl* (Breslau); reprinted 1934 (Marcus, Breslau) and in *Frege 1950*.
  1891 *Function und Begriff. Vortrag gehalten in der Sitzung vom 9. Januar 1891 der Jenaischen Gesellschaft für Medicin und Naturwissenschaft* (Jena); reprinted in *Frege 1962*, 16–37; English translation in *Frege 1952*, 21–41.
  1892 Ueber Begriff und Gegenstand, *Vierteljahrschrift für Wissenschaftliche Philosophie 16*, 192–205; reprinted in *Frege 1962*, 64–78; English translation in *Frege 1952*, 42–55.
  1892a Über Sinn und Bedeutung, *Zeitschrift für Philosophie und philosophische Kritik*, new series, *100*, 25–50; reprinted in *Frege 1962*, 38–63; English translation in *Frege 1952*, 56–78.
  1893 *Grundgesetze der Arithmetik, begriffsschriftlich abgeleitet* (Jena), vol. 1; reprinted 1962 (Olms, Hildesheim); partial English translation in *Frege 1964a*.

1896 Ueber die Begriffsschrift des Herrn Peano und meine eigene, *Berichte über die Verhandlungen der Königlich Sächsischen Gesellschaft der Wissenschaften zu Leipzig, Mathematisch-physikalische Klasse 48*, 361–378.

1903 *Grundgesetze der Arithmetik, begriffsschriftlich abgeleitet* (Pohle, Jena), vol. 2; reprinted 1962 (Olms, Hildesheim); partial English translation in *Frege 1964a*.

1950 *The foundations of arithmetic, A logico-mathematical enquiry into the concept of number*, English translation of *Frege 1884*, with the German text, by John Langshaw Austin (Blackwell, Oxford; Philosophical Library, New York); 2nd revised ed., 1953; reprinted, but without the German text, 1960 (Harper, New York).

1952 *Translations from the philosophical writings of Gottlob Frege*, edited by Peter Geach and Max Black (Blackwell, Oxford); 2nd ed., 1960.

1962 *Funktion, Begriff, Bedeutung. Fünf logische Studien*, edited by Günther Patzig (Vandenhoeck and Ruprecht, Göttingen); 2nd ed., revised, 1965.

1964 *Begriffsschrift und andere Aufsätze*, edited by Ignacio Angelelli (Olms, Hildesheim); contains reprints of *Frege 1879, 1879a, 1882*, and *1882a*; see *Angelelli and Bynum 1966*.

1964a *The basic laws of arithmetic. Exposition of the system*, translated and edited, with an introduction, by Montgomery Furth (University of California Press, Berkeley and Los Angeles, California).

Gödel, Kurt

1930 Über die Vollständigkeit des Logikkalküls (doctoral dissertation, University of Vienna).

1930a Die Vollständigkeit der Axiome des logischen Funktionenkalküls, *Monatshefte für Mathematik und Physik 37*, 349–360; English translation in *van Heijenoort 1967*, 582–591.

1930b Einige metamathematische Resultate über Entscheidungsdefinitheit und Widerspruchsfreiheit, *Anzeiger der Akademie der Wissenschaften in Wien, Mathematisch-naturwissenschaftliche Klasse 67*, 214–215 (communicated on 23 October 1930 by Hans Hahn); English translation in *van Heijenoort 1967*, 595–596; 86–87 of the present volume.

1931 Über formal unentscheidbare Sätze der Principia mathematica und verwandter System I, *Monatshefte für Mathematik und Physik 38*, 173–198; English translations in *Gödel 1962*, 35–72 (but see *Bauer-Mengelberg 1965*), in *Davis 1965*, 5–38 (but see *Bauer-Mengelberg 1966*), and in *van Heijenoort 1967*, 596–616; 87–107 of the present volume.

1931a Über Vollständigkeit und Widerspruchsfreiheit, *Ergebnisse eines mathematischen Kolloquiums 3* (1932), 12–13; English translation in *van Heijenoort 1967*, 616–617; 107–108 of the present volume.

1934 *On undecidable propositions of formal mathematical systems*, lecture notes by Stephen Cole Kleene and John Barkley Rosser (The Institute for Advanced Study, Princeton, New Jersey); reprinted with corrections, emendations, and a postscript in *Davis 1965*, 39–74.

1934a Über die Länge von Beweisen, *Ergebnisse eines mathematischen Kolloquiums 7* (1936), 23–24; English translation in *Davis 1965*, 82–83.

1958 Über eine bisher noch nicht benützte Erweiterung des finiten Standpunktes, *Dialectica 12*, 280–287.

1962 *On formally undecidable propositions of Principia mathematica and related systems*, English translation of *Gödel 1931* by B. Meltzer, with an introduction by R. B. Braithwaite (Oliver and Boyd, Edinburgh and London).

REFERENCES

Herbrand, Jacques
- *1930* *Recherches sur la théorie de la démonstration,* doctoral dissertation, University of Paris; also *Prace Towarzystwa Naukowego Warsawskiego, Wydzial III,* no. 33; partial English translation in *van Heijenoort 1967,* 525–581.
- *1931* Sur le problème fondamental de la logique mathématique, *Sprawozdania z posiedzeń Towarzystwa Naukowego Warszawskiego, Wydzial III,* 24, 12–56.
- *1931a* Unsigned note on *Herbrand 1930, Annales de l'Université de Paris* 6, 186–189.
- *1931b* Sur la non-contradiction de l'arithmétique, *Journal für die reine und angewandte Mathematik* 166, 1–8; English translation in *van Heijenoort 1967,* 618–628.

Hilbert, David
- *1922* Neubegründung der Mathematik (Erste Mitteilung), *Abhandlungen aus dem mathematischen Seminar der Hamburgischen Universität* 1, 157–177; reprinted in *Hilbert 1935,* 157–177.
- *1922a* Die logischen Grundlagen der Mathematik, *Mathematische Annalen* 88 (1923), 151–165; reprinted in *Hilbert 1935,* 178–191.
- *1925* Über das Unendliche, *Mathematische Annalen* 95 (1926), 161–190; reprinted in abbreviated form as *Hilbert 1925a* and also in abbreviated form, but with minor emendations and additions, in *Hilbert 1930a,* 262–288; partial English translation in *Benacerraf and Putnam 1964,* 134–151; English translation in *van Heijenoort 1967,* 367–392.
- *1925a* Über das Unendliche, *Jahresbericht der Deutschen Mathematiker-Vereinigung* 36 (1927), 1st section, 201–215.
- *1927* Die Grundlagen der Mathematik, *Abhandlungen aus dem mathematischen Seminar der Hamburgischen Universität* 6 (1928), 65–85; reprinted in *Hilbert 1928,* 1–21, and *1930a,* 289–312; English translation in *van Heijenoort 1967,* 464–479.
- *1928* *Die Grundlagen der Mathematik,* mit Zusätzen von Hermann Weyl und Paul Bernays, Hamburger Mathematische Einzelschriften 5 (Teubner, Leipzig).
- *1928a* Probleme der Grundlegung der Mathematik, *Atti del Congresso internazionale dei matematici, Bologna 3–10 settembre 1928* (Zanichelli, Bologna, 1929), vol. 1, 135–141; reprinted, with emendations and additions, in *Mathematische Annalen* 102 (1929), 1–9, and *Hilbert 1930a,* 313–323.
- *1930* Die Grundlegung der elementaren Zahlenlehre, *Mathematische Annalen* 104 (1931), 485–494; reprinted in part in *Hilbert 1935,* 192–195.
- *1930a* *Grundlagen der Geometrie* (Teubner, Leipzig and Berlin), 7th ed.
- *1935* *Gesammelte Abhandlungen* (Springer, Berlin), vol. 3.

—— and Wilhelm Ackermann
- *1928* *Grundzüge der theoretischen Logik* (Springer, Berlin).
- *1938* —— 2nd ed.

—— and Paul Bernays
- *1934* *Grundlagen der Mathematik* (Springer, Berlin), vol. 1.
- *1939* —— vol. 2.

Jourdain, Philip Edward Bertrand
- *1912* The development of the theories of mathematical logic and the principles of mathematics, *The quarterly journal of pure and applied mathematics* 43, 219–314.

Kleene, Stephen Cole
- *1936* General recursive functions of natural numbers, *Mathematische Annalen* 112, 727–742; for an erratum and a simplification see *Kleene 1938,* footnote 4, *Péter 1937a,* and *The journal of symbolic logic* 4, iv; for an addendum see *Davis 1965,* 253; reprinted in *Davis 1965,* 237–253.

REFERENCES

Lukasiewicz, Jan
- *1930* Untersuchungen über den Aussagenkalkul, *Sprawozdania z posiedzeń Towarzystwa Naukowego Warszawskiego, Wydział III*, **23**, 30–50; English translation in *Tarski 1956*, 38–59.

Peano, Guiseppe
- *1894* Notations de logique mathématique. Introduction au formulaire de mathématique (Turin).

Princeton University
- *1946* *Problems of mathematics*, Princeton University bicentennial conferences, series 2, conference 2.

Putnam, Hilary
- See Benacerraf, Paul, and Hilary Putnam.

Quine, Willard Van Orman
- *1940* *Mathematical logic* (Harvard University Press, Cambridge, Massachusetts); revised edition 1951; reprinted 1962 (Harper, New York).

Rosser, John Barkley
- *1936* Extensions of some theorems of Gödel and Church, *The journal of symbolic logic* **1**, 87–91; reprinted in *Davis 1965*, 231–235.

Russell, Bertrand
- *1902* Letter to Frege, in *van Heijenoort 1967*, 124–125.
- *1908* Mr. Haldane on infinity, *Mind*, new series, **17**, 238–242.
- See Whitehead, Alfred North, and Bertrand Russell.

Schröder, Ernst
- *1880* Review of *Frege 1879*, *Zeitschrift für Mathematik und Physik* **25**, Historisch literarische Abtheilung, 81–94.

Skolem, Thoralf
- *1923* Begründung der elementaren Arithmetik durch die rekurrierende Denkweise ohne Anwendung scheinbarer Veränderlichen mit unendlichem Ausdehnungsbereich, *Videnskapsselskapets skrifter, I. Matematisk-naturvidenskabelig klasse*, no. 6; English translation in *van Heijenoort 1967*, 302–333.

Tarski, Alfred
- *1953* *Undecidable theories*, in collaboration with Andrzej Mostowski and Raphael M. Robinson (North-Holland, Amsterdam).
- *1956* *Logic, semantics, metamathematics* (Oxford University Press, Oxford).

Trendelenburg, Adolf
- *1867* *Historische Beiträge zur Philosophie* (Berlin), vol. 3, Vermischte Abhandlungen.

Turing, Alan Mathison
- *1937* On computable numbers, with an application to the Entscheidungsproblem, *Proceedings of the London Mathematical Society*, 2nd series, **42**, 230–265; correction, ibid. **43**, 544–546; reprinted in *Davis 1965*, 116–154.

van Heijenoort, Jean (ed.)
- *1967* *From Frege to Gödel: A source book in mathematical logic, 1879–1931* (Harvard University Press, Cambridge, Massachusetts).

von Neumann, John
- *1925* Eine Axiomatisierung der Mengenlehre, *Journal für die reine und angewandte Mathematik* **154**, 219–240; Berichtigung, ibid. **155**, 128; reprinted in *von Neumann 1961*, 34–56; English translation in *van Heijenoort 1967*, 393–413.
- *1927* Zur Hilbertschen Beweistheorie, *Mathematische Zeitschrift* **26**, 1–46; reprinted in *von Neuman 1961*, 256–300.

1928 Über die Definition durch transfinite Induktion und verwandte Fragen der allgemeinen Mengenlehre, *Mathematische Annalen 99*, 373–391; reprinted in *von Neumann 1961*, 320–338.

1929 Über eine Widerspruchsfreiheitsfrage in der axiomatischen Mengenlehre, *Journal für die reine und angewandte Mathematik 160*, 227–241; reprinted in *von Neumann 1961*, 494–508.

1961 *Collected works* (Pergamon Press, New York), vol. 1.

Weyl, Hermann

1949 *Philosophy of mathematics and natural science* (Princeton University Press, Princeton, New Jersey).

Whitehead, Alfred North, and Bertrand Russell

1910 *Principia mathematica* (Cambridge University Press, Cambridge, England), vol. 1.

1912 ——— vol. 2.

1913 ——— vol. 3.

1925 ——— 2nd ed., vol. 1.

1927 ——— vol. 2.

1927a ——— vol. 3.

Index

Ackermann, Wilhelm, 93, 103
Axiom of choice, 84

Bauer-Mengelberg, Stefan, 5, 86
Bernays, Paul, 85, 87
Bernoulli, Jakob, 62
Boole, George, 1

Consistency, 85, 86, 105–108
Continuum problem, 84

Dedekind, Richard, 4, 84
Diaz, Carmen, 1

Fraenkel, Abraham A., 87
Frege, Gottlob, 1–4, 10–12, 26, 66

Gödel, Kurt, 83–87, 103, 106

Herbrand, Jacques, 83, 85
Hilbert, David, 84–87, 93, 101, 103, 107

Jourdain, Philip Edward Bertrand, 1, 4, 10–12, 26

Kant, Immanuel, 21, 55
Kleene, Stephen Cole, 85

Mathematical induction, 4, 62

Paradoxes, 3; the Liar, 83, 89; Richard's, 83, 89
Patzig, Günther, 1
Peano, Giuseppe, 3, 90

Quine, Willard Van Orman, 4

Richard, Jules, 83, 89
Rosser, John Barkley, 85
Russell, Bertrand, 3, 4, 10, 87

Schröder, Ernst, 2, 15
Sequence, logical definition of, 1, 55–81
Skolem, Thoralf, 84

Tarski, Alfred, 85
Trendelenburg, Adolf, 1
Turing, Alan Mathison, 85, 107

von Neumann, John, 87, 91, 92, 96

Lightning Source UK Ltd.
Milton Keynes UK
UKOW042109220911

179138UK00004B/29/A